APPLICATION OF NMR TECHNIQUES ON THE BODY COMPOSITION OF LIVE ANIMALS

A seminar organised by the Commission of the European Communities in the framework of the Community programme for the Coordination of Agricultural Research and held at the Institut für Tierzucht und Tierverhalten, Mariensee (FRG), 14–15 June 1988.

APPLICATION OF NMR TECHNIQUES ON THE BODY COMPOSITION OF LIVE ANIMALS

Edited by

E. KALLWEIT, M. HENNING

Institut für Tierzucht und Tierverhalten FAL, Mariensee, Neustadt, Federal Republic of Germany

and

E. GROENEVELD

University of Illinois, Urbana, Illinois, USA

ELSEVIER APPLIED SCIENCE
LONDON and NEW YORK

ELSEVIER SCIENCE PUBLISHERS LTD
Crown House, Linton Road, Barking, Essex IG11 8JU, England

Sole Distributor in the USA and Canada
ELSEVIER SCIENCE PUBLISHING CO., INC.
655 Avenue of the Americas, New York, NY 10010, USA

WITH 21 TABLES AND 79 ILLUSTRATIONS

British Library Cataloguing in Publication Data

Application of NMR techniques on the body composition
of live animals.
1. Livestock. Biochemistry. Nuclear magnetic
resonance spectroscopy
I. Kallweit, E. II. Henning, M. III. Groeneveld, E.
636.089'2015028

ISBN 1-85166-404-1

Library of Congress CIP data applied for

Publication arrangements by Commission of the European Communities, Directorate-General Telecommunications, Information Industries and Innovation, Scientific and Technical Communication Unit, Luxembourg

EUR 11713

Printed in Great Britain by Galliard (Printers) Ltd, Great Yarmouth

PREFACE

The seminar on "The application of NMR techniques on the body composition of live animals", was held on 14 - 15 June, 1988 at the Bundesforschungsanstalt für Landwirtschaft (FAL), Institut für Tierzucht und Tierverhalten, Mariensee, Germany. This was the third in a series of meetings organized in the framework of the European Communities (EC) programme for the coordination of agricultural research. The earlier meetings were on "In vivo estimation of body composition", in 1981 in Denmark, and "In vivo measurements of body composition in meat animals", in 1983 in the UK..

The emphasis of the Mariensee meeting on the application of nuclear magnetic resonance techniques benefitted greatly from new equipment just installed at the Institute. Furthermore it was made clear that this type of large-scale facility, which is not readily found elsewhere, could be made available to promote international collaboration in animal science. After discussing the basic principles of the NMR technique and comparing it with alternative, non-destructive methods of body composition analyses, the programme of the Mariensee group was discussed. The obvious advantages of the NMR technique were acknowledged and will clearly have to be weighed against the rather high investment costs for the equipment at present.

The Commission of the European Communities acknowledges the major contributions made by the participants, the expert guidance of the chairmen of the sessions, and the particular efforts by the local organizers to create the right working atmosphere to foster Community cooperation and maximise the use of excellent new modern facilities recently constructed at Mariensee.

CONTENTS

SESSION I

PRINCIPLES OF NMR TECHNIQUES

Chairperson: J. Frahm

Principles of NMR - Techniques.

A. Ganssen, Erlangen, FRG

INTRODUCTION

When E. M. Purcell, H. C. Torrey and R. V. Pound as well as F. Bloch, W. W. Hansen, and M. Packard could demonstrate nuclear magnetic resonance in condensed matter for the first time in the winter 1945/1946 (1.) (2.), they would by no means have expected that this finding would lead to so many important applications also in biology and medicine.

One of the reasons for the success of NMR - applications in medicine, biology and related sciences might be the fact, that this method, which can provide information non destructively on the atomic and molecular distribution and dynamics within condensed matter, is most sensitive with respect to the light elements within the periodic system - especially hydrogen which plays an important role in all living organisms.

Considering the abundance of elements in living matter (3.) and comparing it with the relative sensitivity of NMR with respect to the corresponding detectable isotopes (Fig. 1 and Fig. 2)

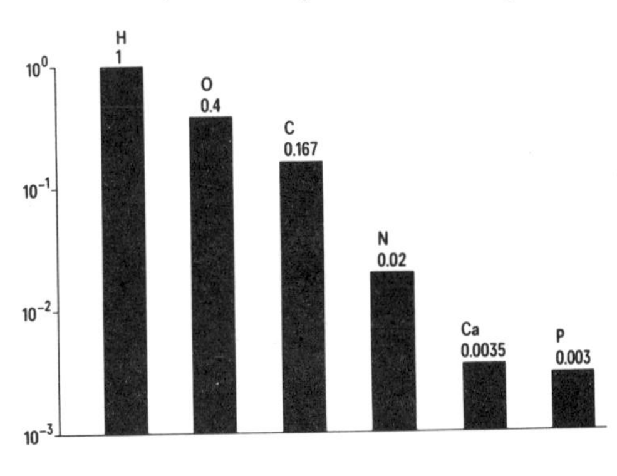

Fig. 1. Relative abundance of biologically important elements in the human body.

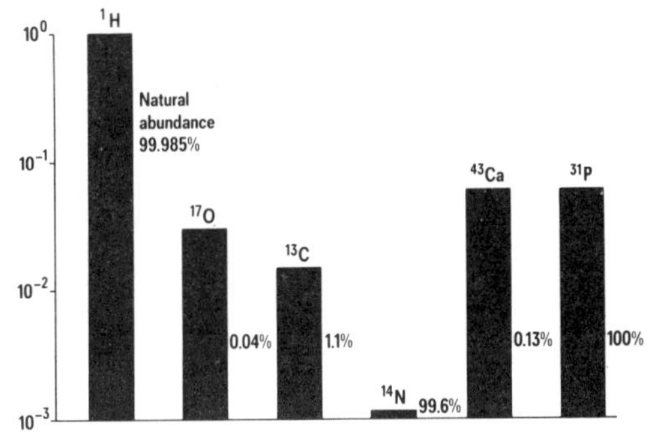

Fig. 2. Relative NMR-sensitivity for some important elements contained within the human body.

we recognize that besides hydrogen atoms or protons other biologically important stable isotopes can be investigated in natural abundance. Of special interest are the isotopes carbon 13, and phosphorus 31.

Another reason for the choice of NMR for the nondestructive study of body composition is the fact that in this technique the medium used for the information transfer through body tissue consists of radiofrequency, electromagnetic waves which can penetrate the body very well. On Fig. 3 the penetration depth of the electromagnetic waves with respect to the human body is plotted as function of frequency. Two windows can be recognized. One is the well known X-ray window with all it's advantages and drawbacks of the high energy ionizing radiation involved. The other window opens up at the lower energy end of the electromagnetic spectrum in the lower microwave region. Within the r. f. range normally used for NMR-imaging hardly any absorption effects can be observed in living subjects. Fig. 3

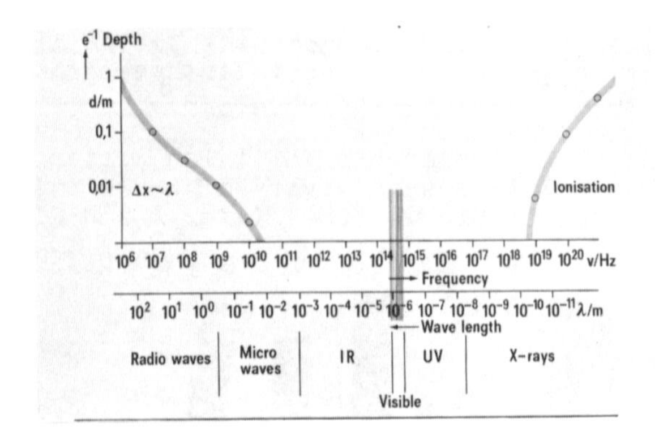

Fig. 3. Penetration depth of electromagnetic waves as function of frequency (wavelength).

NUCLEAR MAGNETIC RESONANCE

Basicly all atomic nuclei containing an odd number of protons or neutrons - or an odd number of both - can be detected with NMR. This means that all these nuclei posses a mechanical angular momentum called spin which gives rise, to a magnetic moment, because of the rotating electric charge distribution within the nucleus.

When such a nuclear magnetic dipole μ_o is being exposed to a magnetic field (B_o) it can occupy a discrete number of magnetic energy states. In the case of the proton and many other nuclei with the spin moment 1/2 (Fig. 4).

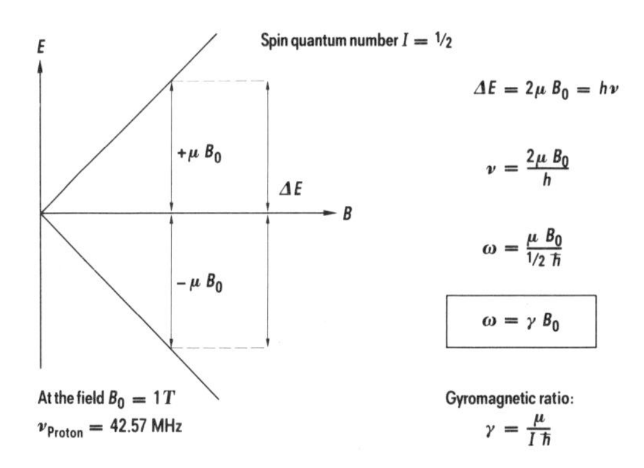

Fig. 4. Magnetic energy of nucleus with the spinquantum number I = 1/2 as function of the magnetic field B.

there are two energy states - spin up or spin down with respect to the magnetic field - with the energy difference:

$$\Delta E = 2 \mu_o B_o$$

According to the rules of quantum mechanics any transition between these two energy states is connected with the absorption or emission of quanta $h\nu_o$, where h is Planck's constant and ν_o the frequency of the electromagnetic wave connected with the transition. This is expressed in the relation:

$$\Delta E = 2 \mu_o B_o = h\nu_o$$

ν_o is also called the nuclear magnetic resonance frequency, which is for protons in a field

$$B_o = 1 \text{ T} : \nu_o = 42.57 \text{ MHz}$$

Since the radiation connected with this magnetic transition is circularly polarized usually the angular frequency $\omega_o = \nu_o 2\pi$ is being used in the resonance equation:

$$\omega_o = \gamma B_o$$

where the gyromagnetic ratio γ characteristic for the nuclear isotope represents the ratio between the magnetic moment of the nucleus, and the mechanical spin momentum.

Nuclear magnetic properties, natural isotopic abundances and relative NMR - sensitivities of some stable isotopes which are important for it's application on living organisms are listed in Fig. 5.

Nucleus	γ MHz/T	Natural Isotopic Abundance	Relative Sensitivity*	Spin
^{1}H	42.576	99.985	1	1/2
^{2}H	6.536	0.015	0.0096	1
^{13}C	10.705	1.108	0.016	1/2
^{14}N	3.076	99.635	0.001	1
^{15}N	4.315	0.365	0.001	1/2
^{17}O	5.772	0.037	0.029	3/2
^{19}F	40.055	100	0.834	1/2
^{23}Na	11.262	100	0.093	3/2
^{31}P	17.236	100	0.066	1/2
^{33}S	3.266	0.74	0.0023	3/2
^{39}K	1.987	93.08	0.0005	3/2

*At constant field, with sensitivity of the 1-hydrogen nucleus = 1.

Fig. 5. Data important for the NMR application of stable isotopes, which are contained in living matter.

Let us consider an assembly of magnetic dipoles or spins in a liquid or even solid body Fig. 6.

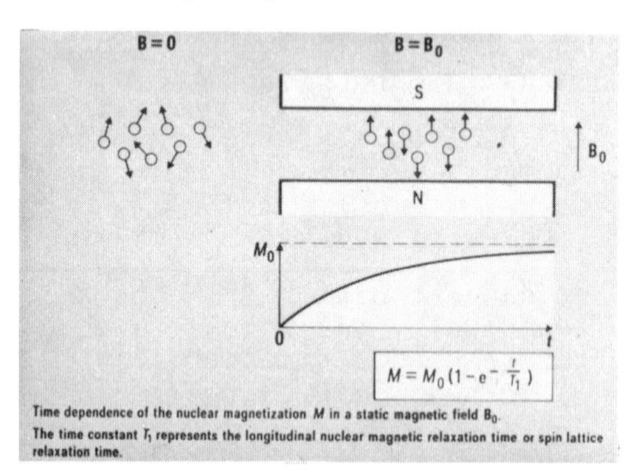

Time dependence of the nuclear magnetization M in a static magnetic field B_0.
The time constant T_1 represents the longitudinal nuclear magnetic relaxation time or spin lattice relaxation time.

Fig. 6. Nuclear magnetization in a static magnetic field. The longitudinal relaxation time T_1 determins the time dependence of the longitudinal nuclear magnetization.

At room - or body temperature they are fluctuating relative to each other corresponding to their thermal energy while the spins are randomly oriented. On application of a magnetic field one half of the spins will be adjusted parallel, the other half antiparallel.

However by interacting with each other they will assume exponentially an equilibrium magnetization. This means that a part of the spins will assume a lower magnetic energy state so that there will be a net magnetic moment of the sample, which is determined by the Boltzman - factor, the ratio of magnetic energy versus thermal energy.

At the magnetic field of 1 T for instance the excess population in the lower energy state is very small, and only 10^{-6} of the total number of spins. The time constant for the magnetization is called the <u>*spin-lattice relaxation time T_1.*</u>

Macroscopically the assembly of spins within for instance a volume of water can be treated as one magnetic dipole which can exhibit a magnetic dipole momentum given by the excess population of the magnetic energy states. This dipole momentum is also called the nuclear magnetization. When on resonance - this means when an electromagnetic field perpendicular to the constant field with the resonance frequency $\omega_0 = \quad B_0$ is being applied - the system will behave similarly like a spinning top within the gravitational field of the earth.

The magnetization will be tipped over from the allignment with B_0 and gyrate around the main field direction B_0 with the resonance frequency ω_0.

The basic scheme of a pulse-type NMR - apparatus is showm on Fig. 7.

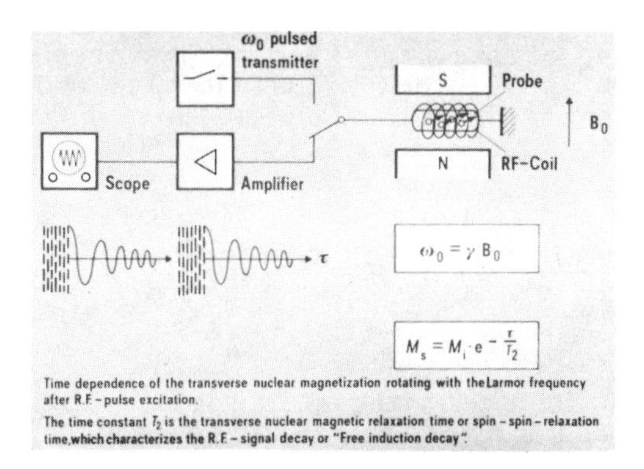

Time dependence of the transverse nuclear magnetization rotating with the Larmor frequency after R.F.-pulse excitation.
The time constant T_2 is the transverse nuclear magnetic relaxation time or spin-spin-relaxation time, which characterizes the R.F.-signal decay or "Free induction decay".

Fig. 7. The NMR-experiment. Stimulation of NMR-signal by r. f. pulse with the frequency ω_0. Free nuclear induction decay (FID) of the induced signal with time constant T_2.

The sample to be investigated is exposed to a constant magnetic field B_o within a magnet. Simultaniously a r. f. pulse with the field directron perpendicular to the constant field can be applied to the sample thereby creating a rotating (precessing) magnetic moment. By actuating a switch the same r. f. coil can be connected to a r. f. -amplifier and the amplified voltage picked up by the r. f. coil can then be displayed on an oscilloscope. There will be a response in form of an exponentially decaying r. f. signal, the socalled free induction decay (FID). The time constant for the decay is being called the transversal or spin-spin relaxation time T_2.

In a practical magnetic field which is usually not ideally homogeneous the signal decays much faster. The spins lose their phase coherence due to the inhomogenious field distribution. (Fig. 8)

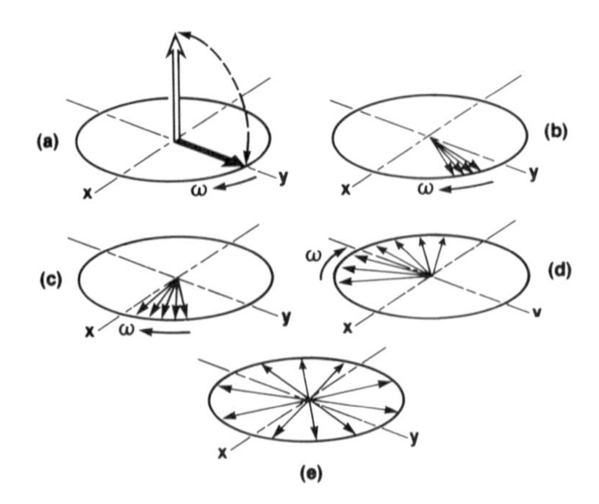

Fig. 8. Dephasing of the precessing transverse magnetization after 90° pulse. NMR signal becomes undetectable for random phase distribution (e.)

To remove this influence of magnetic field inhomogeneity at least temporarily, Carr, and Purcell proposed (4.) already in the early fifties, the creation of "spin echoes". After an *exciting r. f. pulse which turns the magnetization by 90° from the main magnetic field direction a 180° r. f. pulse of double length is being applied after a time interval TI. This way phase coherence can be produced at least for a short time again after a time interval of 2 TI called a "spin echo" Fig. 9.*

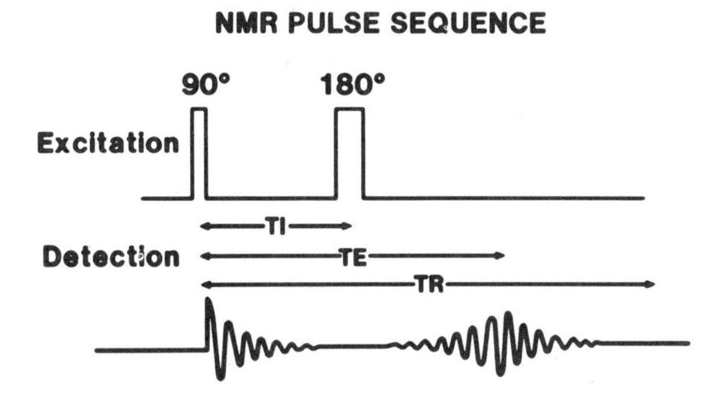

Fig. 9. NMR-pulse sequence for the creation of a spin echo. By 180^{0} inversion of transverse magnetization after TI the spin dephasing caused by d. c. field inhomogeneity can shortly be annihilated after 2 TI.

From the exponential decrease of the spin echo amplitudes with time the true relaxation time T_2 can be derived.

This can be done by the application of a series of 90°, and repeated 180° pulses, a so called "multi echo series" (Fig. 10).

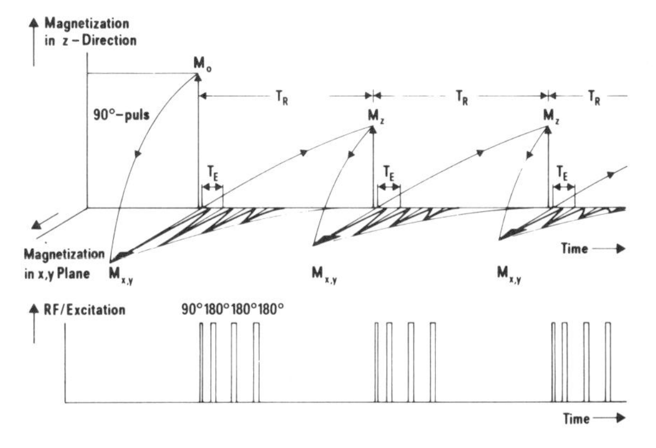

Fig. 10. Nuclear magnetization in multiple spin echo series.

Starting from an equilibrium magnetization first a 90° pulse will be applied to obtain a detectable signal. By successive 180° pulses an echo series can be produced for the evaluation of T_2. The time interval between the first 90° pulse and the echo is called the "echo time" T_E, and the time between successive pulse series the "repetition time" T_R. A considerable number of different NMR pulse series has been developed in the mean time.

Fig 11 gives examples of some simpler pulse series which became also important for NMR - imaging.

Some examples of measured relaxation times T_1 and T_2 within a life rat (5.) are given in Fig. 12.

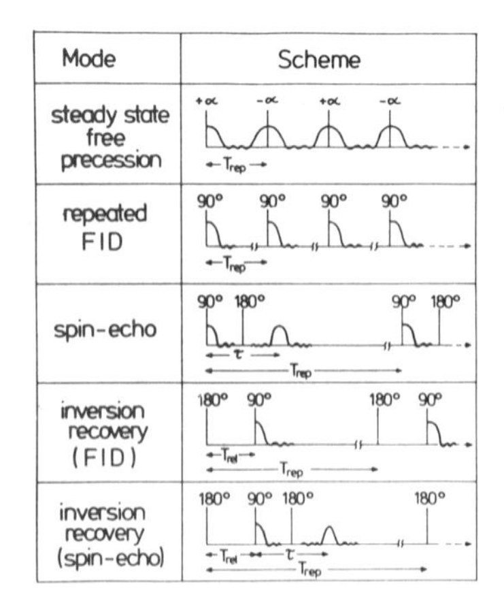

Fig. 11. Examples of some simpler NMR-pulse series.

Since the tissues are never homogeneous the relaxation times are always effective relaxation times composed of a considerable number of different contributions of the different tissue components. Because the relaxation times are strongly dependant on molecular mobility the longer relaxation times are usually *found in tissues with a high content of unbound water. An increasing abundance of higher molecular weight compounds usually decreases the relaxation times - so do paramagnetic ions by increasing the magnetic interaction with their neighbouring nuclear magnetic dipoles. Paramagnetic ions, like for instance some rare earth ions, are therefore being used as contrast agents in NMR - imaging.*

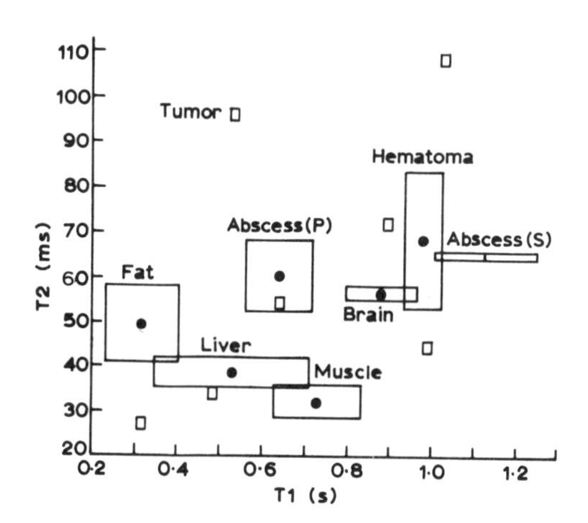

Fig. 12. Nuclear magnetic relaxation times T_1 and T_2 of different tissue in rat.

NMR-IMAGING:

Most of the NMR-investigations are performed as analytical procedures in vitro on physically and chemically homogeneous material. For the NMR-investigation of highly inhomogeneous living organisms like men or animals we need spatial resolution between different volume elements possibly exhibiting different NMR-properties within much more extended bodies. For special purposes NMR-surface coil focussing on regions of interest or field focussing can be a solution to this problem. (6.) (7.) (8.). More general and for the biological and medical application extremely fertile was the idea of P. Lauterbur in 1973 for a NMR-imaging scheme which he called "MR-Zeugmatography". (9.).

He proposed the application of linear magnetic field gradients superimposed on the d. c.-magnetic field, so that spatial location in gradient direction is being transformed into a corresponding NMR-resonance frequency Fig. 13.

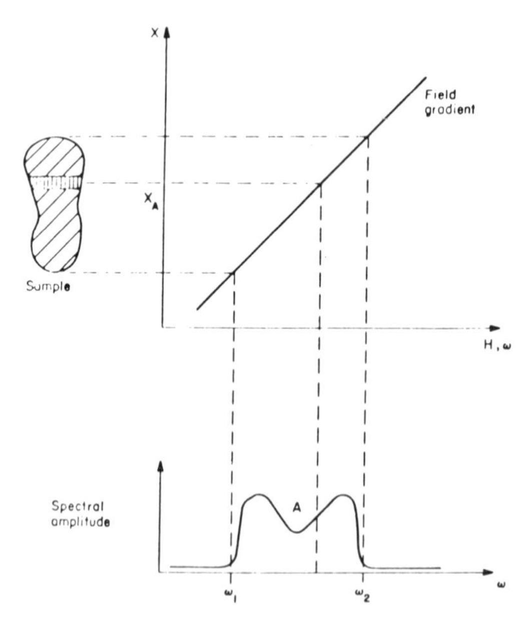

Fig. 13. Conversion of the proton density distribution of a sample within a magnetic field into a frequency distribution with the help of a linear, magnetic field gradient and NMR.

We speak of a frequency encoding of the spatial coordinates. This way projections of the spin density within the object under investigation can be obtained, by back conversion from fequency space into geometric space. Analogue to X-ray computer tomography a reconstruction from projections can be performed. For the inital selection of the slice to be imaged, a linear field gradient perpendicular to the desired slice has to be applied while spin excitation is being performed with a r. f. pulse of the corresponding frequency. Fig. 14.

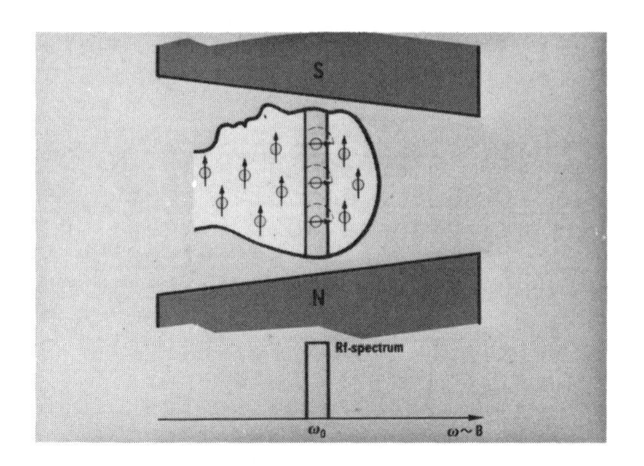

Fig. 14. Selective excitation of a slice.

The projection signal can then be obtained with a gradient perpendicular to the first one with the resonance frequency given by the spatial location with respect to gradient direction and the signal amplitude proportional to the projected spin density. Fig. 15.

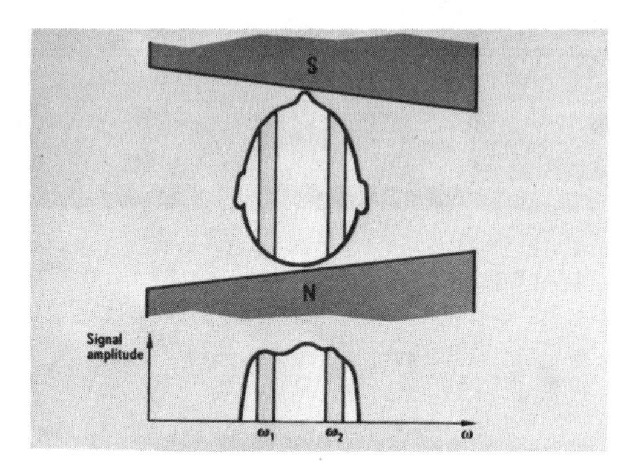

Fig. 15. Recording of a projection in a linear magnetic field gradient.

A sufficient number of projections has to obtained while the projection gradient is being turned incrementely by 360° around the axis at right angles to the imaged plane to be able to reconstract the slice with the desired resolution. In contrast to X-ray computer tomography in NMR-imaging there are various methods for image data collection, and reconstruction.

The now most often applied NMR-Fourier-Zeugmatography which was proposed by Kumar, Welti and Ernst in 1975 (10.) employs in addition to the spatial frequency encoding of the NMR signal also a phase incoding in the orthogonal direction. The diagram Fig. 16 shows the corresponding r. f. and gradient-pulse sequence for two dimensional Fourier-NMR-imaging. After slice selection by excitation in presence of a slice selection gradient a phase encoding gradient is applied at right angles to the original gradient, and it's amplitude (or duration) is successively increased, step by step - instead of progressively rotating the projection gradient in the original Lauterbur method. After the phase encoding period the signal frequency determining, orthogonal projection gradient is turned on during the echo signal. The phase- and frequency encoded signals are collected in increasing order of magnitude of the phase encoding gradient. The final imaging can then be obtained by a two dimensional Fourier transformation of the resulting signal matrix.

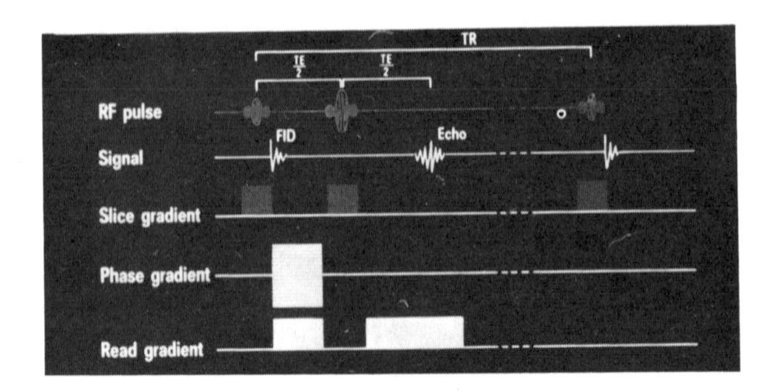

Fig. 16. Spin echo pulse sequence for Fourier-NMR-imaging.

$$S \sim P e^{-TE/T_2} (1 - e^{-TR/T_1})$$

Pure T_1 Image:	Measurement of each voxel with two different TR times and iterative T_1 calculation by computer
Pure T_2 Image:	Measurement of two or more echoes and T_2 calculation by computer
Pure P Image:	Measurement of two or more echoes of protons in a state of complete relaxation (very long TR) and extrapolation by computer

Fig. 17. Time dependence of the spin echo signal in NMR-imaging.

A 3 dimensional image matrix can be obtained in a similar way. The relation for the signal obtained with the spin echo method, most often used in NMR-imaging is given in table Fig. 17.

The importance of the right choice of the sequence parameters T_R and T_E is demonstrated in Fig. 18.

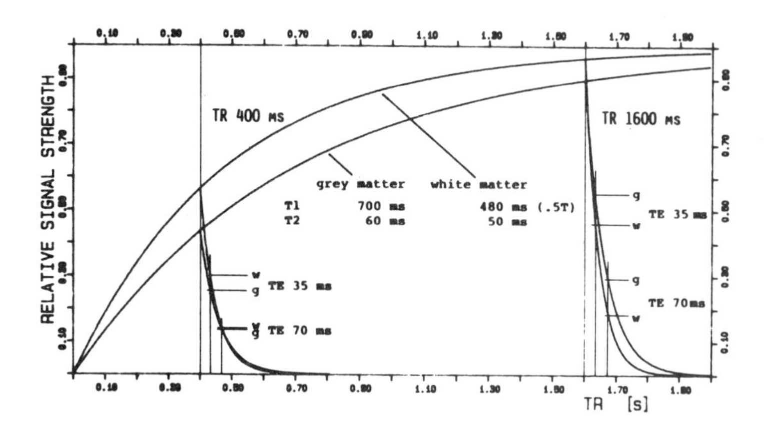

Fig. 18. Influence of repetition time T_R and echo delay TE on contrast of grey and white matter of human brain.

As example of the time dependence of the proton magnetization of gray- and white matter tissue of the human brain is plotted for two different repetition times T_R and echo times T_E. It becomes evident that the biggest contrast between the two types of tissue evolves with the longer T_R-value of 1.6 s and the longer T_E-value of 70 ms.

To complete the list of parameters which can be displayed with NMR imaging means, it has to be mentioned that also liquid flow, molecular diffusion, NMR-line shift and magnetic field distribution within the scanned plane can presently be obtained with adequate pulse sequences, and more or less data processing.

NMR - SPECTROSCOPY - IN VIVO

In the early 1950's it was recognized already (11.) (12.) that the NMR - resonance frequency varied not only from nucleus to nucleus; there is also a slight deviation of the resonance frequency in dependence of the molecular conformation, and the particular chemical environment at the molecular site carrying the resonating nuclei. It is this property, which is called the chemical shift, which is the base for NMR-spectroscopy. The frequency deviation caused by the slightly different local magnetic fields, the nuclei are exposed to, is usually very small. The chemical shift is being measured by the relative frequency deviation

$$\delta = \frac{\nu_i - \nu_i'}{\nu_i} \quad \text{in parts per million (ppm)}$$

While the chemical shift for proton compounds is only in the order of 0 - 12 ppm, it can be bigger for 31 p compounds about 0 - 30 ppm, and much bigger for 13 C compounds with about 0 - 300 ppm. Some examples for more common chemical radicals, and the reference compounds for 0 ppm deviation are given in table. Fig. 19.

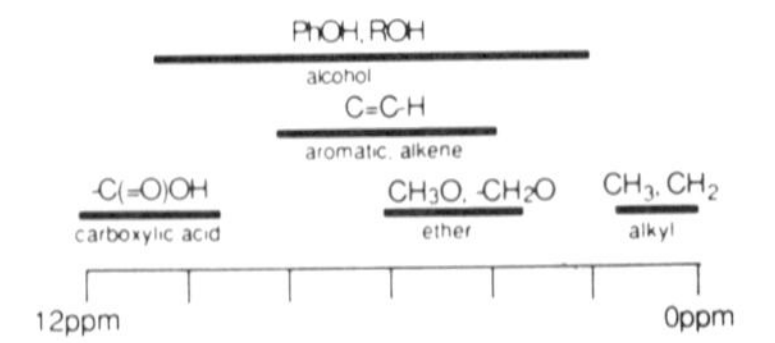

1-Hydrogen chemical shifts of selected functional groups.
Reference compound is tetramethylsilane (TMS).

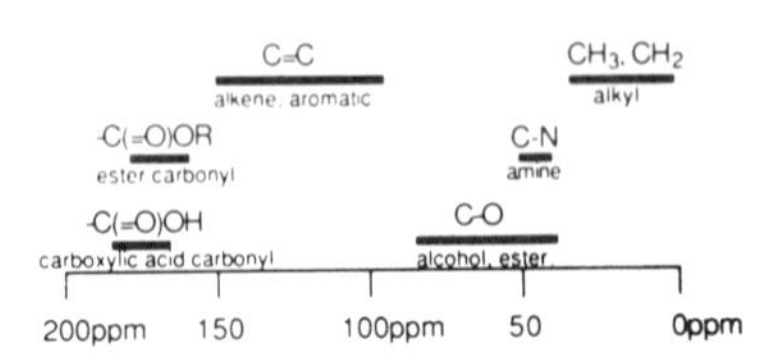

13-Carbon chemical shifts of selected functional groups.
Reference standard is tetramethylsilane (TMS).

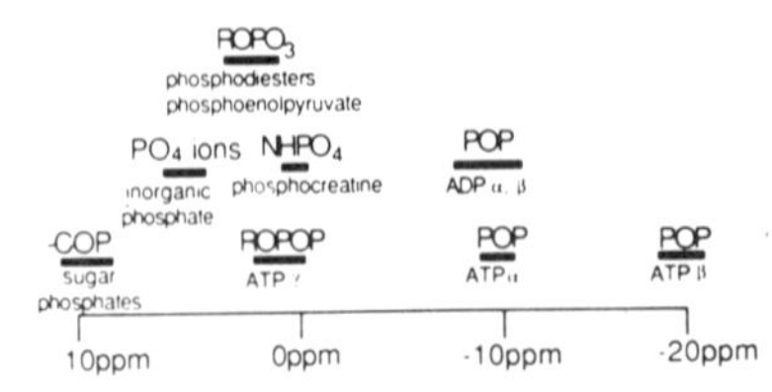

31-Phosphorus chemical shifts of biologically important compounds.
Reference standard is phosphoric acid.

Fig. 19. Chemical shift of different molecular groups for H 1, C 13. and P 31.

The chemical shift of many compounds changes with dilution and sometimes also with the pH-value of the solution.

A serious difficulty for the "in vivo-spectroscopy" is caused by the relatively low concentrations of the more interresting metabolytes, which is in the order of micro-Moles. The diagram Fig. 20

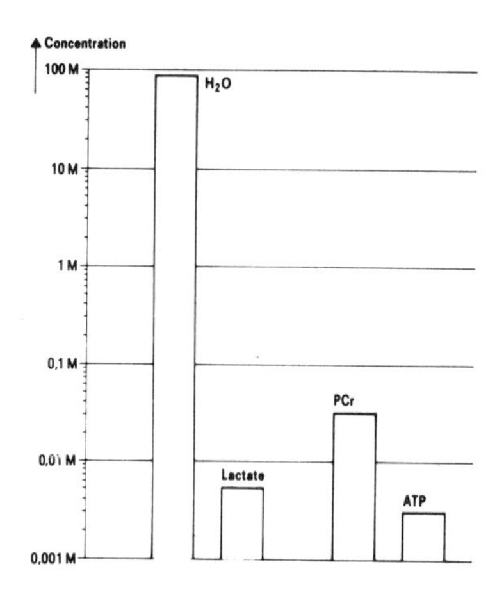

Fig. 20. Average concentrations of some important metabolytes.

shows a comparison between the average concentrations of water, and the more prominent metabolytes like lactate, phosphocreatin, and adenosin-tri-phosphate (ATP). To obtain a well resolved NMR spectrum it is favorable to use the highest possible magnetic field strength - usually at least 1.5 T. This has the additional advantage that the absolute NMR-line resolution increases also, provided that the magnetic field homogeneity is being maintained. In most cases it is necessary to excite a much bigger volume within the tissue of interest as for instance in imaging. While in NMR-imaging, where usually protons in water are being detected often only few mm³ are sufficient for a well detectable signal, in NMR-spetroscopy even at a field strength of more than

1.5 T several ml are usually required to obtain a well resolved in vivo-spectrum within a reasonable time of several minutes. The simplest approach might be the use of a surface coil with a sensitive volume of several ml Fig. 21.

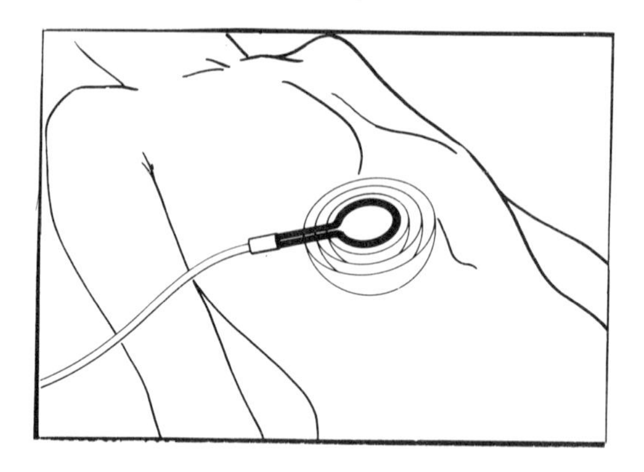

Fig. 21. Surface coil. Measurement volume indicated.

A well defined and limited sensitive volume can be obtained by several methods. The simplest method consists of the effective exclusion of the volume area close to the skin with an r. f. excitation amplitude thus big that optimum detection can only be achieved at greater distance from the coil ("rotating frame method"). (13.)

A better selection of a single slice within the body can be obtained by the additional application a field gradient at right angles to the surface during excitation. With this Depth Resolved Surface Coil Spectroscopy (DRESS) the spectroscopic volume is limited to a disc shaped area within the body. (14.)

A number of combined r. f. and gradient pulse sequences similar to imaging sequences with the excitation confinement to one volume element only have been created for localized volume spectroscopy. In cases of sufficient isotopic concentration like in the case of protons in water and fat it is possible to create spatially resolved images of either of the two comparatively strong, and broad spectral lines, about 4 ppm apart.

A more detailed proton spectroscopy in soft tissue becomes possible only if the rather strong and comparatively broad water line is being reduced by selective saturation, usually with a strong r. f. pulse on the frequency of the water peak.

Most metabolic information has thus far been obtained with ^{31}P - MR spectroscopy. The important metabolytes of the energy metabolism, like phosphocreatine (PCr), adenosine-tri-phosphate (ATP), shugarphosphate (PS), and inorganic phosphates can directly be measured Fig. 22.

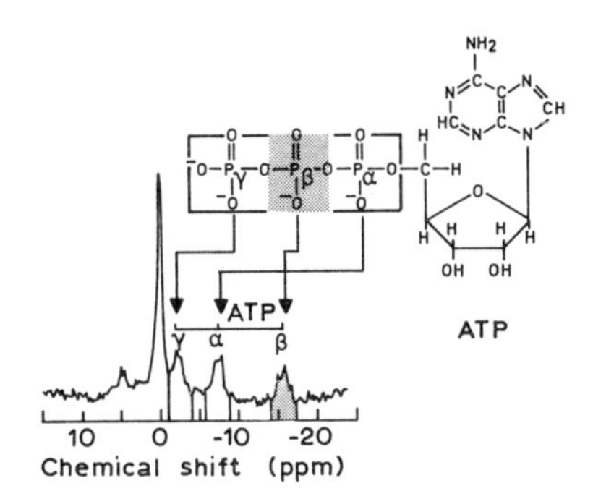

Fig. 22. Chemical shift of the P 31-spectrum of P Cr, ATP, and Pi.

The intracellular pH-value shows up in the line shift between the inorganic phosphate line and the ATP-lines.

The decrease of phosphocreatine and the increase in inorganic phosphates within a human muscle after exercise can be seen in Fig. 23.

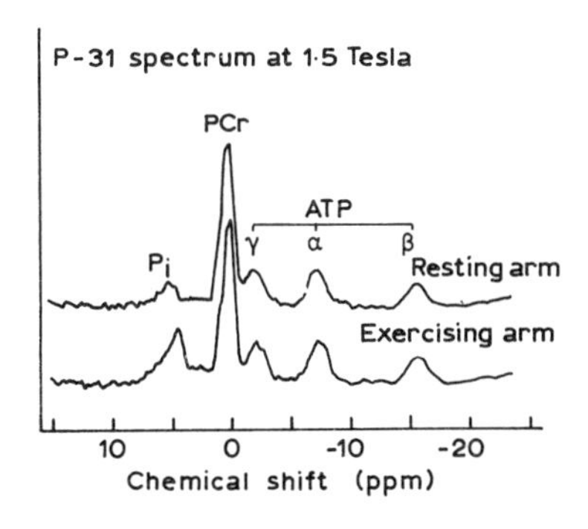

Fig. 23. The influence of physical exercise on the P 31-spectrum in human arm.

The excellen t NMR-spectroscopic resolution achievable with a 4 T field with natural abundance -^{13}C spectroscopy is shown in Fig. 24.

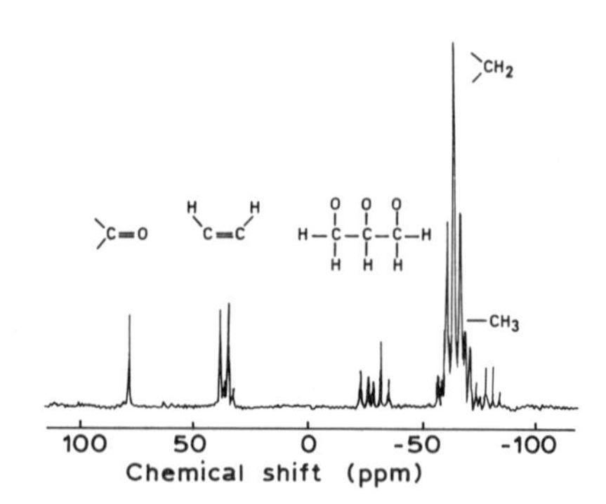

Fig. 24. C 13-spectrum of human adipose tissue at 4 T.

This spectrum was obtained in vivo on adipose tissue with a scantime of 4 min on a localized volume of a few ml.

The previous results of the increasing number of NMR-spectroscopic studies on life animals and men are so encouraging that a considerable number of new applications of NMR-spectroscopy can be expected in the future.

CONCLUSION:

Nuclear magnetic resonance is a modality capable to measure in vivo, and non invasively body composition - as far as the better NMR-detectable elements like hydrogen, carbon, phosphorus, and sodium within their more abundant lower molecular compounds are concerned. Via magnetic relaxation time measurements information can be obtained on molecular mobilities, and diffusion constants. NMR-imaging can provide anatomic information in all three dimensions, as well as on blood flow. NMR-spectroscopy enables us to study body metabolism by detection of the more abundant body metabolytes.

REFERENCES:

Purcell, E. M., Torrey, H. C., and Pound, R. V. 1946, Phys. Rev. 69, 37

Bloch, F., Hansen, W. W., and Packard, M. 1946, Phys. Rev. 70, 474

Documenta Geigy, Wissenschaftliche Tabellen, 6. Auflage, 486

Carr, H. Y., and Purcell, E. M. 1954, Phys. Rev. 94, 630

Davis, P. L., Kaufman, L., Crooks, L. E., 1983, Clinical Magn. Res. Imaging, ed. A. R. Margulis et al., San Francisco

Ganssen, A., 1967, DBP 1566, 148

Ackerman, J. J. H., Grove, T. H., Wong, G. G., Gadian, D. G., Radda G. K., 1980, Nature 283, 167 - 170.

Gordon, R. E., Hanley, Shaw, D. et al., 1980, Nature 287, 736 - 738

Lauterbur, P. C., 1973, Nature 242, 190

Kumar, A. D., Welti, R. R., Ernst, R. R., 1975, Naturwiss. 62, 34.

Proctor, W. G., and Yu, F. C., 1950, Phys. Rev., 77, 717

Arnold, J. T., Dharmatti, S. S., and Packard, M. E., 1951, J. Chem. Phys. 19, 507.

Hoult, D. I., 1979, J. Magn. Res. 33, 183.

Bottomley, P. A., Foster T. B., and Darrow, R. D., 1984, J. Magn. Res. 59, 338

SESSION II

APPLICATION IN HUMAN MEDICINE

Chairperson: A. Ganssen

APPLICATION OF MR IMAGING TECHNIQUES IN HUMAN MEDICINE

P. Heintz, Ch. Ehrenheim, H. Hundeshagen
Department of Nuclear Medicine
Medical School of Hannover
D-3000 Hannover 61

INTRODUCTION

The last ten years have seen a revolution in diagnostic imaging. During this brief period of time, computed tomography and ultrasound have become established as indispensable diagnostic tools, and magnetic resonance imaging has begun to play an important role in clinical diagnosis (HIGGINS and HRICAK, 1987; LISSNER and SEIDERER, 1987; STARK and BRADLEY, 1988).

MRI is a new diagnostic procedure with limited availability. Thus, the following factors should influence its selection as a diagnostic tool in imaging of the human body. This includes the type of information required (sensitivity or specificity), the anatomic location of the suspected abnormality and its suspected nature, the presence of additional abnormalities or significant related conditions, the minimum level of quality and accuracy required of the examination, the ease of obtaining the desired information, the speed with which the results can be obtained, the relevance of the information expected to patient management, and finally the cost-effectiveness (MERRITT, 1985).

The complexity of MR imaging, in both operation and interpretation, poses a real challenge to the diagnostic physician. On the one hand, images are based on several tissue specific parameters including proton density and the relaxation times T1 and T2. On the other hand, we have to select from a large number of operator-controlled parameters in conducting each examination. This includes the type of radiofrequency coil (body coil, high resolution surface coils), the field of view, the imaging planes (direct multiplanar imaging without reformatting), section thickness and section gap, the total number of sections, the pulse sequences (spin echo, bipolar, multi echo techniques, FLASH and FISP), the repetition time (TR), the echo time (TE), and finally the number of data averages. The importance of using proper signal acquisition techniques cannot be overstated, for lesions can go undetected if the setting is inappropriate (HEIKEN et al., 1986).

NMR SIGNAL INTENSITY IN THE BODY

One characteristic of MRI is the excellent soft tissue contrast it provides. Fatty tissue and bone marrow, predominantly consisting of fat, appear very bright in any pulse sequence, and air as well as compact bone are always dark (fig. 1- 3). Thus, calcifications are not seen well. Blood vessels may give a low signal depending on flow (LENZ et al., 1988). Water and other fluids provide a low signal in T1-weighted (dark) and a high signal in T2-weighted images (bright). Tumors are of various signal behaviour.

In order to gain higher sensitivity and specificity paramagnetic contrast agents are used. By shortening mainly the T1 relaxation time of the protons, Gadolinium-DTPA improves the contrast in T1-weighted images. Depending on the perfusion tumors will appear brighter. The delineation of internal details of the lesion is improved (fig. 4).

	T1	T2
FAT	very bright	very bright
AIR	very dark	very dark
BLOOD VESSELS (RAPID FLOW)	very dark	very dark
COMPACT BONE	very dark	very dark
BONE MARROW	very bright	very bright
WATER	very dark	very bright
MUSCLE	mod dark	mod dark
LIGAMENTS AND TENDONS	dark	dark
TUMOR	mod dark	bright

Fig. 1 NMR signal intensity in the body (MERRITT, 1985).

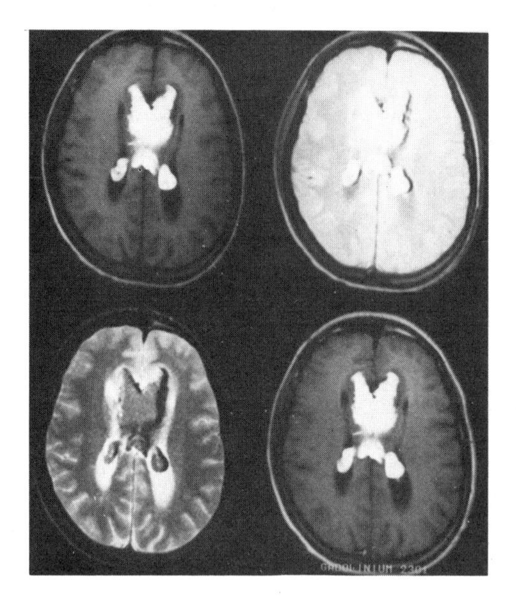

Fig. 2 Transverse section of the head. The tumor of the corpus callosum has the same signal intensity as the subcutaneous fat and turned out to be a lipoma. It appears bright in T1-weighted (upper left), proton-density- (upper right) and T2-weighted (lower left) images, there is no enhancement after the application of a contrast agent. Proton-density- and T2-weighted image additionally depict a chemical shift artifact as a dark streak in the left of the tumor.

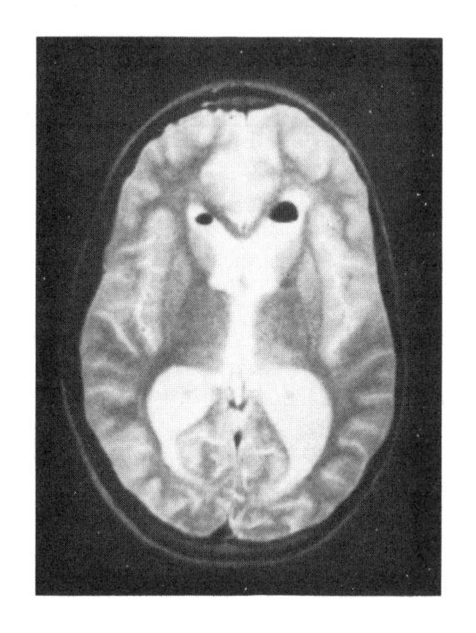

Fig. 3 Half-rounded dark area in all pulse sequences (T2-weighted image shown) within each ventricornu, while CSF forms a level. This is caused by air which remained in the ventricular system after surgical treatment.

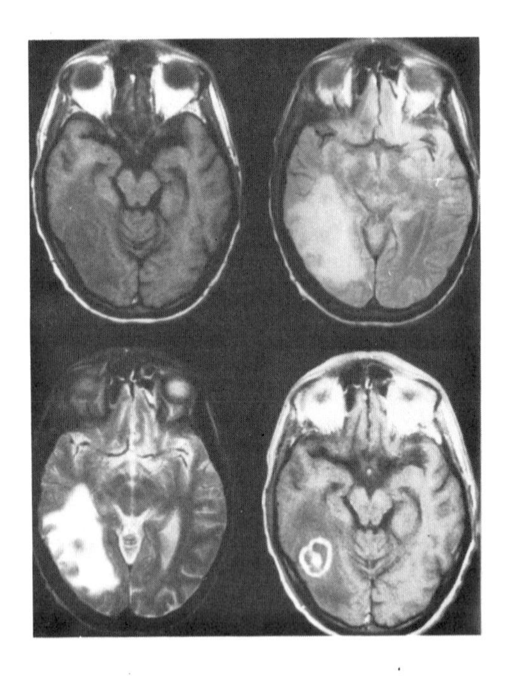

Fig. 4 Paramagnetic contrast agents improve especially the demarcation of lesion and surrounding edema. In this case of acquired immune deficiency syndrome the T2-weighted image (lower left) reveals the extent of the edema best, whereas the inflammatory changes of toxoplasmosis can be identified by an annular contrast enhancement after intravenous application of Gadolinium-DTPA (lower right).

PROBLEMS OF MR IMAGING AND ARTIFACTS

A schematic diagram of the upper abdomen in transverse section shows the motion artifacts of this area (fig. 5). The ghosts of moving structures such as subcutaneous fat, aorta, heart, and stomach propagate vertically outside the patient along the phase-encoding axis.

Exposures which are not synchronized to the cardiac phase are prone to strong signal reductions and the blurred reconstruction of moving structures. These influences are reduced by ECG triggering which uses the R wave as reference for starting measurements (fig. 6). A freely selectable delay time can be chosen to measure various stages of the cardiac cycle.

The method applied to keep respiratory artifacts to a minimum is shown in figure 7. Respiratory gating means that the measurement system uses only the data measured during the respiratory end phase for image computation. During the respiratory end phase the thorax is not prone to pronounced respiratory movement. The measurement sequence is not interrupted during gating. The measurement time may be twice as long (HEINTZ et al., 1988a). Figure 8 compares exposures without and with respiratory gating.

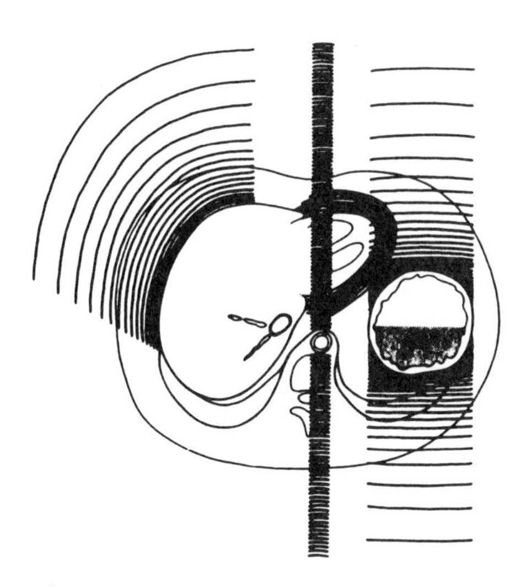

Fig. 5 Motion artifacts of the upper abdomen (from STARK and BRADLEY, 1988).

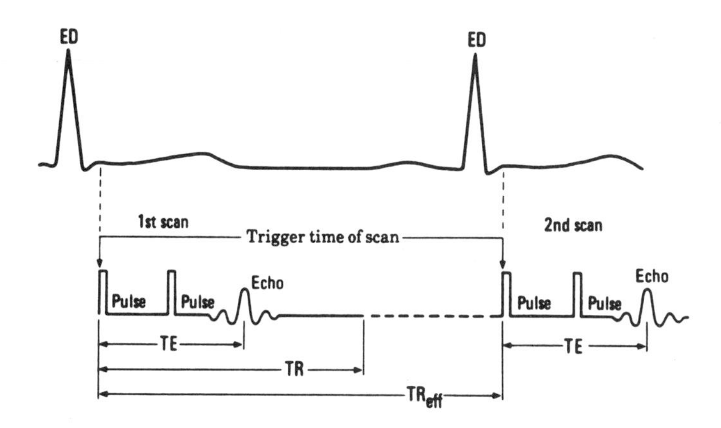

Fig. 6 An ECG triggered image of the heart during a particular phase of the cardiac cycle is generated by taking scans (for each slice) triggered with the same delay time during this phase. The timing parameters of the scans (repetition time TR, echo time TE) must be adjusted to the cardiac cycle of the patient. TR_{eff} = effective repetition time is a function of the RR interval (from SIEMENS, 1987).

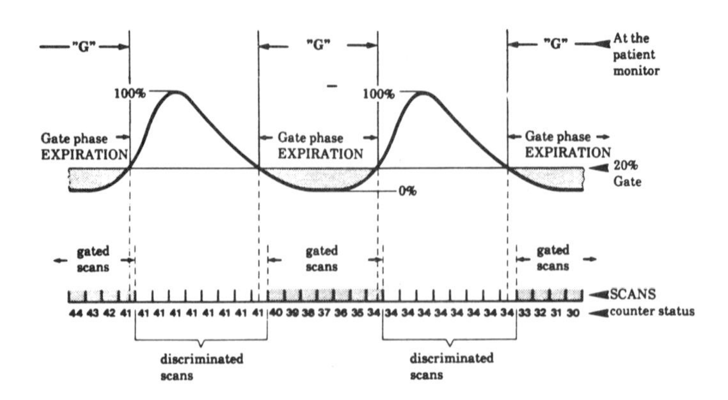

Fig. 7 The respiratory signal displayed on the patient monitor rises during inspiration and drops during expiration. The respiratory signal is used for generating the gate phase signal according to the selected gate threshold (from SIEMENS, 1987).

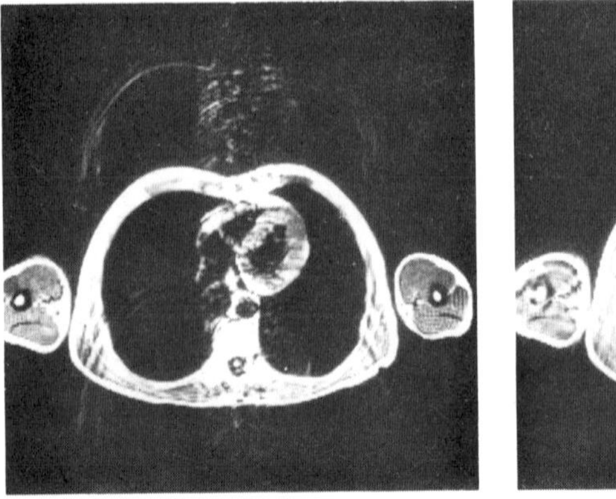
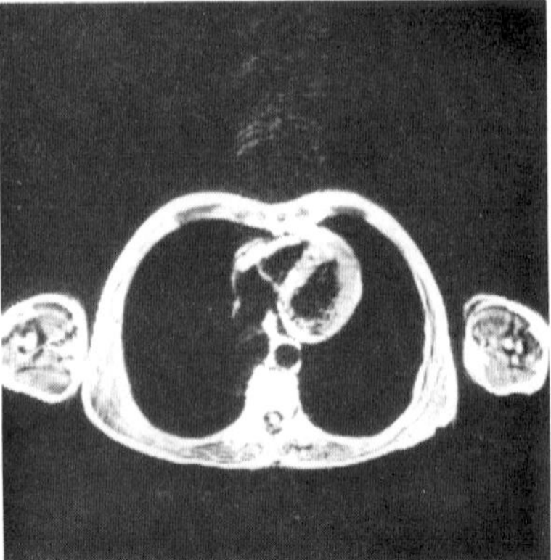

Fig. 8 Exposures without and with respiratory gating. ECG triggering was used in both sequences. There is an impressive reduction of ghosts of motion artifacts on the right. The number of data averages was one in both cases, that means an averaging of two data acquisitions.

An other method of motion artifact reduction without prolongation of the acquisition time by triggering or gating is available now. Conventional spin echo sequences with additional gradients are used for the so called gradient motion refocussing (LENZ et al., 1988). For example, flow artifacts of CSF dorsal of the spinal cord may cause serious problems of interpretation. The use of GMR sequences removes the described intraspinal artifacts in the sagittal plane (fig. 9) as well as in the transverse.

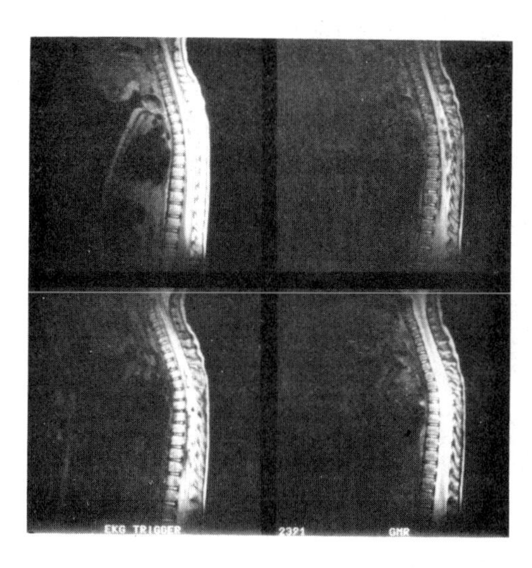

Fig. 9 Reduction of CSF-flow artifacts by use of GMR sequences. T1-weighted image without artifacts. Distinct artifacts appear in the T2-weighted (upper right) and in the ECG triggered T2-weighted (lower left) image. The improvement of the T2-weighted image by use of GMR is obvious (lower right). In this case respiratory movement artifacts appearing as sagittal streaks are reduced, as well.

FUNCTIONAL MR IMAGING

An example for functional MR imaging using fast imaging sequences based on gradient echoes is the examination of the transplanted kidney (HEINTZ et al., 1988b). During intravenous bolus application of Gadolinium-DTPA a FLASH sequence with a flip angle of 30 degrees (T1-weighted) is applied. Sequential scans of a coronal slice are obtained with a time increment of 8 s (minimum 3 s) and increasing delay times. Total examination time is about 10 min p.i. The application of the contrast agent, which is excreted by glomerular filtration only, reveals disturbances of renal perfusion and function (fig 10). The high resolution of this functional imaging method allows the differentiation of various renal structures, such as renal cortex, medulla, collecting system, and infarction. Figure 11 demonstrates the possibility of quantitative assessment of the regional perfusion and function of the kidney by generation of signal intensity versus time curves.

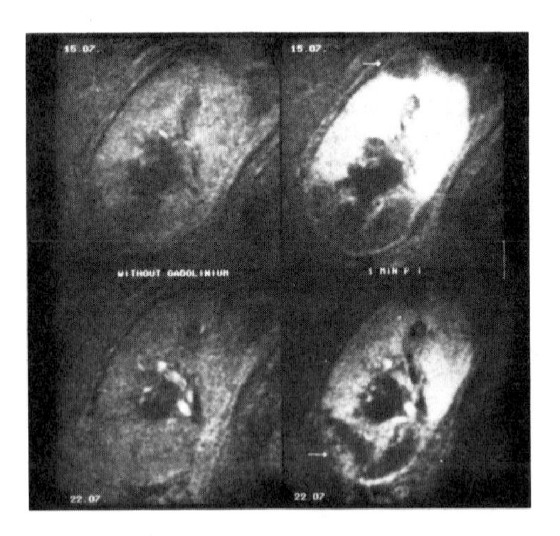

Fig. 10 Exemplary exposures of the dynamic study. At an early stage the lack of perfusion is seen as a defect, whereas one week after the acute infarction the beginning peripheral reperfusion can be detected.

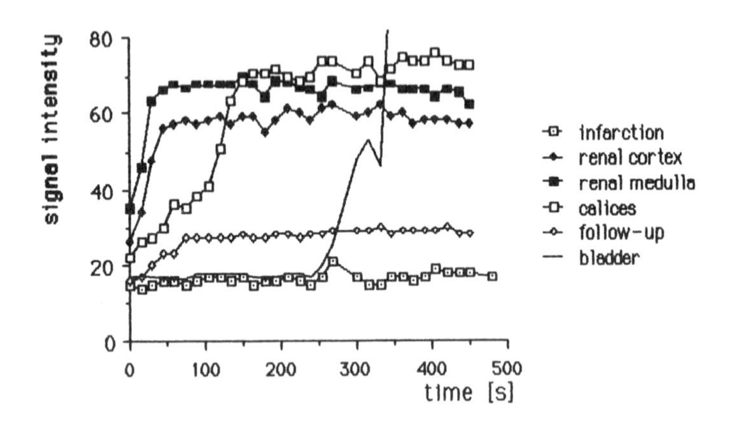

Fig. 11 Signal versus time curves of the dynamic MR study. The signal intensity curve of the infarction is similar to the background curve.

RELAXATION TIMES AND FAT WATER SEPARATION

As mentioned above fat appears bright in all pulse sequences because of high proton density, short T1 and long T2. Cortical bone is dark in any image because of low proton density, long T1 and short T2. The tissue in the middle of the diagram (fig. 12), for example cerebrospinal fluid, appears dark in T1- and bright in T2-weighted images because of long T1 and long T2.

The figures 13 and 14 give examples of tissue differentiation by calculation of relaxation times and use of fat water separation.

brightest (high proton density, short T1, long T2)	darkest (low proton density, long T1, short T2

FAT
NUCLEUS PULPOSUS
BONE MARROW
SPINAL CORD
MUSCLE
CSF
ANNULUS FIBROSUS
LIGAMENTS
CORTICAL BONE

Fig. 12 Schematic diagram of signal intensities and relaxation times in the area of the spine.

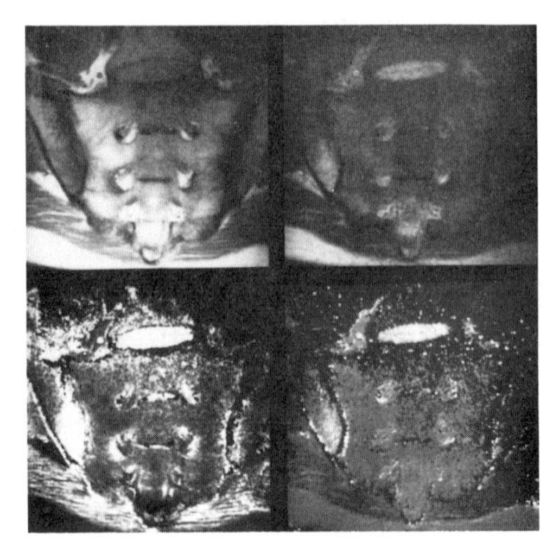

Fig. 13 Paraxial coronal sections of the sacroiliac joints in T1- and T2-weighted pulse sequences (upper row). The T2-weighted images reveal the right sided sacrocoxitis as an area of higher signal intensity. The calculated T1- and T2-weighted images (lower row) confirm the diagnosis and depict the real extension of the inflammatory changes. The water content of the edema is comparable to that of the lumbar disc according to the signal intensity.

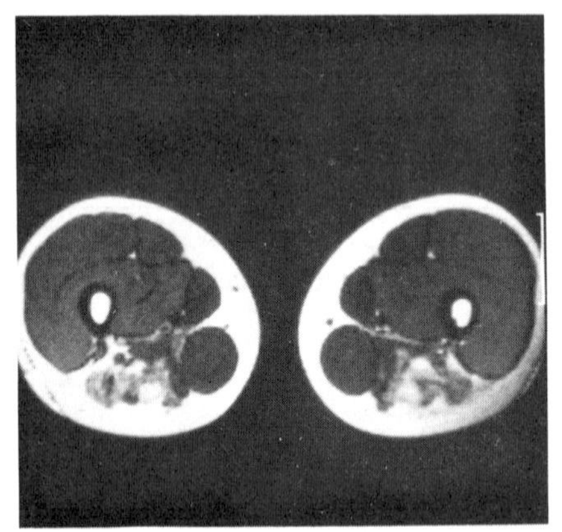

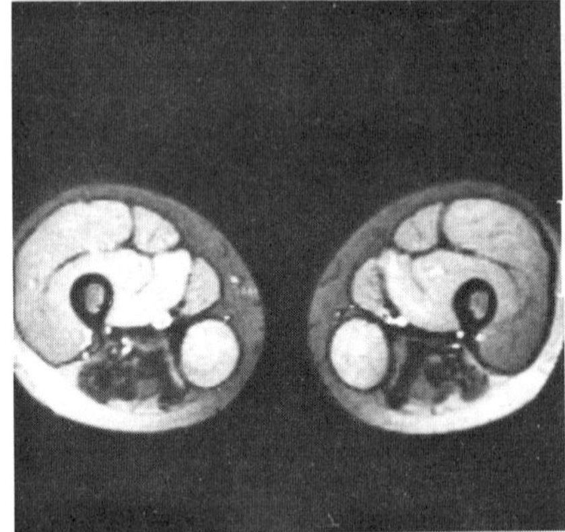

Fig. 14 Extensive symmetrical changes of the dorsal muscles of the thigh. Their signal behaviour (bright in T1-weighted and T2-weighted images, not shown) lead to the presumptive diagnosis of fatty degenerations. The images of the fat water separation confirm the diagnosis. The fat image (on the left) pronounces fatty structures by high signal intensity, whereas the water image pronounces a high water content by high signal intensity.

ADVANTAGES AND DISADVANTAGE

The advantages include the absence of ionizing radiation, generally the absence of harmful side effects, the possibility of multiplanar imaging with images based on several tissue specific parameters (proton density, relaxation times) rather than one. MRI provides blood flow information, excellent contrast resolution, good spatial resolution, and the possibility of image processing. Variable pulse sequences are available to optimize the study and paramagnetic contrast agents may improve the fair specificity, while the sensitivity is excellent, at any rate.

The disadvantages include a limited patient throughput, high costs, limitations of slice thickness and limited information about calcifications and compact bone. Pacemakers, vascular clips, metallic implants (fig. 15), and claustrophobia belong to the patients exclusions.

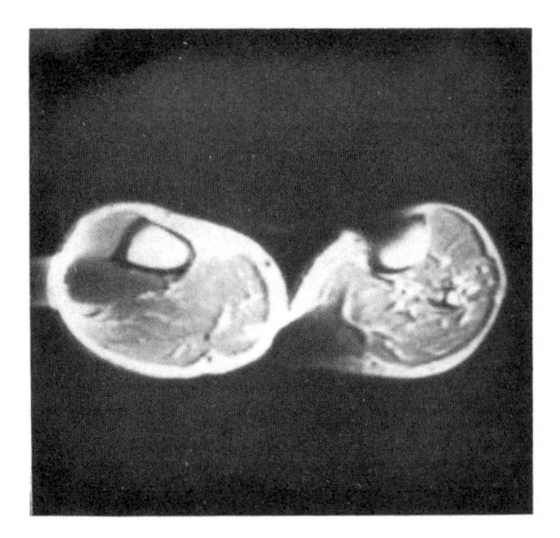

Fig. 15 Multiple metallic splinters in the lower leg of a war-disabled person are the origin of distinct artifacts. They cause considerable distorsion of the contours of the leg, which make an interpretation difficult. Depending on the localization metallic splinters represent an exclusion for patients examination.

REFERENCES

Heiken, J.P., Glazer, H.S., Lee, J.K.T., Murphy, W.A., Gado, M. 1986. Manual of Clinical Magnetic Resonance Imaging. (Raven Press, New York)

Heintz, P., Ehrenheim, Ch., Hundeshagen, H. 1988a. Technik und klinischer Nutzen des Atemgating in der Kernspintomographie. Röntgenpraxis, in press

Heintz, P., Ehrenheim, Ch., Hundeshagen, H. 1988b. Gadolinium-DTPA in MRI and ^{99m}Technetium-DTPA in the Examination of the Perfusion of Transplanted Kidneys. Diagnostic Imaging International, in press

Higgins, Ch. B., Hricak, H. 1987. Magnetic Resonance Imaging of the Body. (Raven Press, New York)

Lenz, G.W., Haacke, E.M., Masaryk, T.J., Laub, G. 1988. In-plane Vascular Imaging: Pulse Sequence Design and Strategy. Radiology, 166, 875-882

Lissner, J., Seiderer, M. 1987. Klinische Kernspintomographie. (Ferdinand Enke, Stuttgart)

Merritt, C.R.B. 1985. Clinical Applications of Magnetic Resonmance Imaging - An Overview. In "Magnetic Resonance Imaging. MRI Training Program for Physicians" (Ochsner Medical Institutions)

Siemens. 1987. ECG triggering and respiratory gating, beginning with Version C1. In "Supplement to the MAGNETOM operating instructions" (Siemens Aktiengesellschaft, Medical Engineering Group, Erlangen)

Stark, D.D., Bradley, W.G. 1988. Magnetic Resonance Imaging. (Mosby, St. Louis, Washington, Toronto)

LOCALIZED PROTON NMR SPECTROSCOPY OF HUMAN BRAIN IN VIVO

Jens Frahm, Harald Bruhn, Michael L. Gyngell, Klaus-Dietmar Merboldt,
Wolfgang Hänicke, and Rolf Sauter*

Max-Planck-Institut für biophysikalische Chemie
Postfach 2841, D-3400 Göttingen, FRG
*Bereich Medizinische Technik, Siemens AG,
Henkestraße 127, D-8520 Erlangen, FRG

SUMMARY

Complementary to anatomical descriptions and functional studies of the human body with high spatial resolution and soft tissue contrast by advanced magnetic resonance imaging (MRI), image-selected localized proton magnetic resonance spectroscopy (MRS) is demonstrated to allow a non-invasive biochemical tissue characterization in vivo. After suppression of the unwanted water signal and focussing onto a desired volume-of-interest (VOI) metabolite resonance signals in localized proton MR spectra represent biochemical fingerprints of normal and pathological tissues. Information is derived on the energy metabolism, amino acids, fatty acids and lipid membrane precursors, neurotransmitters, and secondary messengers. Relaxation times give further access to molecular dynamics, in vivo concentrations, and transport processes of major metabolites. Preliminary patient studies demonstrate an exciting potential for a non-invasive metabolic assessment of cerebral tumors.

INTRODUCTION

The feasibility of localized water-suppressed proton MRS of human brain in vivo has been demonstrated on a conventional 1.5 tesla whole-body MRI system (Siemens Magnetom). Localization is achieved using the stimulated echo (STEAM) spectroscopy sequence (Frahm, 1987a; Gyngell, 1988; Frahm, 1988a; Frahm, 1988b)

$$90^{o}(\text{Slice 1}) - TE/2 - 90^{o}(\text{Slice 2}) - TM - 90^{o}(\text{Slice 3}) - TE/2 - STE$$

comprising three slice-selective 90° rf pulses in the presence of orthogonal magnetic field gradients. Only the second half of the stimulated echo (STE) appearing at the echo time TE is acquired in the absence of any gradient. Fourier transformation of the time-domain data then yields the desired MR spectrum. Since the STE arises only from spins that experience all three rf pulses, the spectrum stems from a limited VOI defined by the intersection of all three slices. Water suppression is performed by preceding chemical-shift-selective 90° rf pulses exciting only water protons. Their magnetizations are spoiled by subsequent applications of magnetic field gradients.

STUDIES OF THE NORMAL HUMAN BRAIN IN VIVO

A standard protocol for the investigation of a volunteer or patient comprises image selection of the VOI using fast scan FLASH MRI (Haase, 1986; Frahm, 1986; Frahm, 1987b; Frahm, 1987c) localized optimization of the magnetic field homogeneity as well as water suppression, and spectroscopic investigations at different echo times and repetition times. Typically, high-resolution proton MR spectra with resonance linewidths of 5 Hz were obtained from 27 ml or 64 ml VOIs within measuring times of 1-10 min. Figure 1 shows a fully relaxed proton MR spectrum of a 64 ml VOI localized in the occipital area of the brain of a normal volunteer. Experimental parameters were TE = 50 ms, TM = 30 ms, TR = 6000 ms, and 128 scans. Resonance assignments

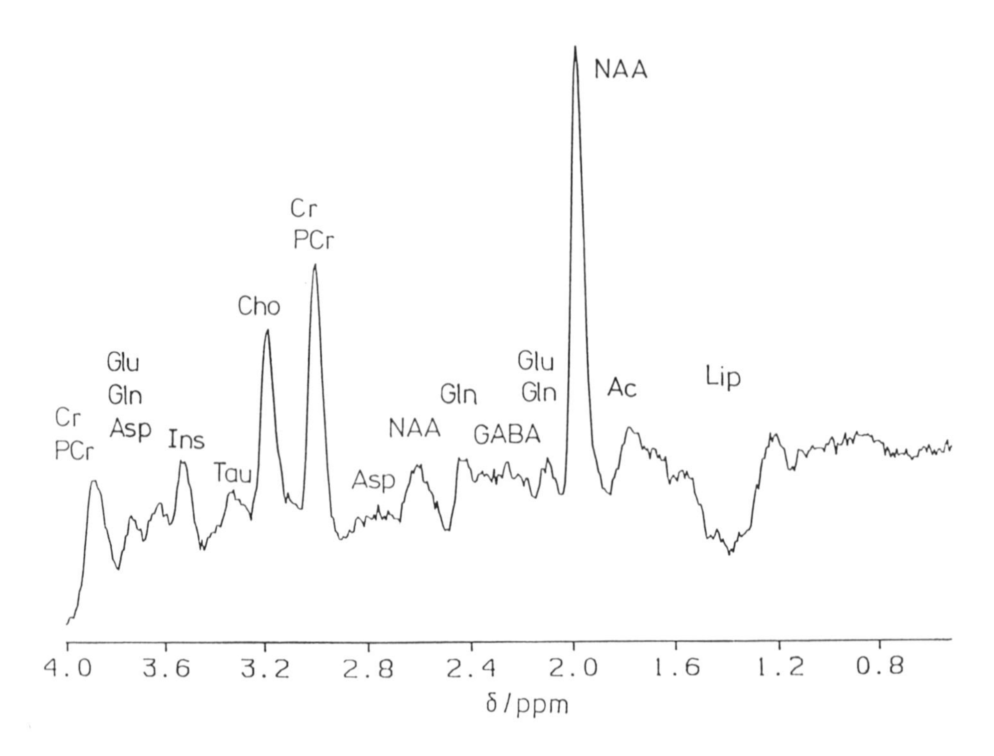

Fig. 1: 64 MHz (1.5 T) water-suppressed localized proton MR spectrum of a normal volunteer. A 64 ml VOI (4 cm x 4 cm x 4 cm) has been localized in the occipital area of the brain. The spectrum was acquired at TE = 50 ms (TM = 30 ms) using 128 scans and TR = 6000 ms corresponding to fully relaxed conditions without T1 saturation. Resonance assignments are made to lipids (Lip), acetate (Ac), γ-amino butyric acid (GABA), N-acetyl aspartate (NAA), glutamine (Gln), glutamate (Glu), aspartate (Asp), creatine (Cr), and phosphocreatine (PCr), choline-containing compounds (Cho), inositol phosphates (Ins), and glycine (Gly).

are made to lipids, lactate, acetate, N-acetyl aspartate, γ-amino butyrate, glutamine, glutamate, aspartate, creatine/phosphocreatine, choline compounds, taurine, and inositols (Frahm, 1988a). T1 and T2 relaxation times were determined to calculate metabolite concentrations relative to reference values. Differences in relaxation times and concentrations of major metabolites were found in grey and white matter, thalamus, and cerebellum (Frahm, 1988b).

PATIENT STUDIES

Reliability and reproducibility of the STEAM spectroscopy technique offers great potential for a biochemical description of metabolic disorders. In fact, preliminary applications of localized proton spectroscopy to patients with cerebral tumors resulted in extremely promising findings: (i) proton MR spectra of tumors exhibit striking differences when compared to spectra of normal brain tissue, and (ii) histologically different tumors appear with characteristic spectra or "fingerprints" (Frahm, 1988c; Bruhn, 1988). For example, while resonances from choline compounds were part of both normal and tumor spectra (though with altered T2 values and concentrations), resonances from creatine/phosphocreatine and N-acetyl aspartate were found to be absent in most tumors. Spectral differences between different types of tumors occur in the concentrations of lactate and adenine nucleotides, in the lipid content, and in further unidentified resonances.

CONCLUSION

A full patient study including both MRI and proton MRS can be performed within a total investigational time of 1-1.5 hours. For proton spectroscopy echo times of 50 ms and 270 ms and repetition times of 1.5 s were found to be most useful. Since the results were obtained on a conventional 1.5 T MRI system without special modifications, localized proton MRS may become a routine clinical tool. Currently, applications are extended to other types of pathologies as well as to other organs. Non-invasive metabolic characterizations by localized in vivo NMR bear future potential of replacing needle biopsies.

ACKNOWLEDGEMENT

Financial support by the Bundesminister für Forschung und Technologie (BMFT) of the Federal Republic of Germany (Grant 01 VF 86066) is gratefully acknowledged.

REFERENCES

Bruhn, H., Frahm, J., Gyngell, M.L., Merboldt, K.D., Hänicke, W., Sauter, R. and Hamburger, C. 1988. Noninvasive Differentiation of Human Cerebral Tumors by Localized Proton NMR Spectroscopy In Vivo. Submitted.

Frahm, J., Haase, A. and Matthaei, D. 1986. Rapid NMR Imaging of Dynamic Processes Using the FLASH Technique. Magn. Reson. Med. 3, 321-327.

Frahm, J,, Merboldt, K.D., and Hänicke, W. 1987a. Localized Proton Spectroscopy Using Stimulated Echoes, J. Magn. Reson. 72, 502-508.

Frahm, J., Merboldt, K.D. and Hänicke, W. 1987b. Transverse Coherence in Rapid FLASH NMR Imaging. J. Magn. Reson. 72, 307-314.

Frahm, J., 1987c. Rapid FLASH NMR Imaging. Naturwiss. 74, 415-422.

Frahm, J., Gyngell, M.L., Bruhn, H., Merboldt, K.D., Hänicke, W. and Sauter, R. 1988a. Localized High-Resolution Proton NMR Spectroscopy Using Stimulated Echoes. Initial Applications to Human Brain In Vivo. Magn. Reson. Med., in press.

Frahm, J., Bruhn, H., Gyngell, M.L., Merboldt, K.D., Hänicke, W. and Sauter, R. 1988b. Localized Proton NMR Spectroscopy in Different Regions of the Human Brain In Vivo. Relaxation Times and Concentrations of Cerebral Metabolites. Magn. Reson. Med., submitted.

Frahm, J., Bruhn, H., Gyngell, M.L., Merboldt, K.D., Hänicke, W. and Sauter, R. 1988c. Localized Proton Spectroscopy of a Primary Brain Tumor In Vivo. Submitted.

Gyngell, M.L., Frahm, J., Merboldt, K.D., Hänicke, W. and Bruhn, H. 1988. Motion Rephasing in Gradient-Localized Spectroscopy. J. Magn. Reson. 77, 596.

Haase, A., Frahm, J., Matthaei, D., Hänicke, W. and Merboldt, K.D. 1986. FLASH Imaging. Rapid NMR Imaging Using Low Flip-Angle Pulses. J. Magn. Reson. 67, 258-266.

NMR SPECTROSCOPY. APPLICATIONS IN HUMAN MEDICINE.

G. Kozak-Reiss

C.C.M.L. - C.N.R.S. UA-1159 - 133, av. de la Résistance - F-92350 Le Plessis-Robinson

ABSTRACT

P31 NMR spectroscopy affords a non invasive method to the study of normal and diseased muscles. Significant differences between the spectra of normal and diseased human muscles were reported in some myopathies (Duchenne dystrophy, Mac Ardle disease, mitochondrial myopathies, ...). The possibility that metabolic disturbances will be linked to contraction abnormalities led to investigate Malignant Hyperthermia susceptible muscles in pigs as in humans. Studies carried out on pig muscles showed acidosis and disturbances of energetic metabolism when sustained contraction were elicited by electrical stimulation in MH susceptible muscles and in the presence of triggering agents (halothane, caffeine, Ca2+ ionophore).

Malignant Hyperthermia (MH) is an acute syndrome occuring in genetically predisposed subjects. Exercise Hyperthermie (EH), often called "heat stroke" is an hyperthermic state induced by sustained muscular exercise beyond the subject's capability. These syndromes can often be lethal or leave invaliding sequelae. A method able to detect non invasively this potential risk would thus allow prevention and be wellcome. Whereas, at rest, no difference are observed among patients and normal subjects, MH and EH patients develop an abnormal P31 spectral configuration during exercise and recovery. The main difference between normal and MHS subjects is a faster and deeper decrease in pHi and a slower return to resting value during recovery. This drop in pHi is still accentuated by ischemia. The results indicate a perturbed energy metabolism and most probably an increased anaerobic glycolysis in EH and MH during exercise.

P31 NMRS is a useful non invasive tool both from the standpoint of ability to study and to detect diseased muscles, and from the standpoint of patient's safety because the measurements, unlike biopsies, are harmless.

INTRODUCTION

NMR spectroscopy affords a non invasive method to the studies of human tissue "in vivo" (13). When examining spectra from leaving tissue, only a limited range of molecules is observed. High-resolution spectra are obtained only from molecules that are in rapid molecular motion. That implies for example that bone and tightly bound phospholipids are not seen. Taking into account of their natural abundance, the most intersting nuclei are H1, C13 and P31. In H1 spectra, signals are obtained from water and fat. For P31 spectra, resonances are observed from phosphate metabolites in the cytoplasm. The direct quantitation of lactic acid production was determined by C13 SRM. (1). Another way is to use H1 SRM and to quantify lactic acid from the integral of its methyl proton resonance. At the present time, the major application of NMR to "in vivo" biochemisty of human muscle concern phosphate metabolism (2, 4, 7).

Because the phosphorylated compounds play a key role in cellular metabolism, changes in their concentration may provide an index concerning the state of the muscle. P31 NMR-S (P31-NMR spectroscopy) allows to follow the kinetics of concentration changes of the main phosphorylated metabolites such as creatine phosphate (CP), inorganic phosphate (Pi), ATP, as well as intracellular pH (pHi).

P31 NMR-S has been used in recent years for studying various myopathies. Malignant Hyperthermia (MH) is an acute syndrome occuring in genetically predisposed subjects. Exercise Hyperthermia (EH), often called "heat stroke" is an hyperthermic state induced by sustained muscular exercise beyond the subject's capability. These syndromes can often be lethal or leave invalidating sequelae. A method able to detect non invasively this potentiel risk would thus allow prevention and be wellcome.

The possibility that metabolic disturbances will be linked to contraction abnormalities led to investigate Malignant Hyperthermia susceptible muscles in humans as in pigs (8, 9, 10, 12). Studies carried out "in vivo" on pig muscles showed acidosis and disturbances of energetic metabolism when sustained contraction were elicited by electrical stimulation (unpublished results). The same for superfused muscle bundles when triggering agents halothane, caffeine and Ca2+ ionophore were added to the physiological media (8). Therefore, Malignant Hyperthermia Susceptible (MHS) subjects underwent NMR-P31 spectroscopy. The same for subjects who suffered from Exercise Hyperthermia (heat stroke) (9, 10).

METHODS

1 - NMR-P31 spectroscopy

NMR spectra were obtained using a Fourier spectrometer interfaced with a 2 Tesla, 55 cm bore, superconducting magnet operating at 34.8 MHz resonant frequency for P31. Data were acquired following radiofrequency pulses at 1.5 sec intervals with a 3 cm diameter surface coil (90° flip angle at the coil's center). Sixty free induction decays (90 s) were summed for each spectrum at rest, during exercise and last part of recovery period, and 40 FIDS (60 s) during the first 5 minutes of recovery.

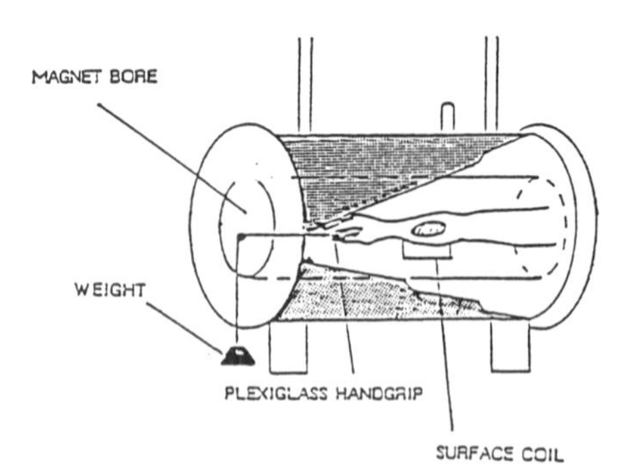

Fig. 1 Experimental device used to study the energetic metabolism of human muscles.

2 - Exercise protocol

The 6 minutes exercise consisted in finger flexion lifting weight (2-4 kg), repeated every 2 seconds. The work done during non ischemic exercise was showed in order to get no more than 50 % decrease in CP for each subject. The same exercise protocol was applied both in normoxic and hypoxic conditions, which were produced by arm compression with a cuff. Normal circulatory conditions were restored at the end of exercise. It was verified that a 6 minutes ischemia without exercise neither affected the phosphate metabolites nor the pHi.

3) Subjects

Ten healthy volunteers were investigated as control subjects.

Eighty four subjects underwent both P31-SRM and muscle biopsy to diagnose MH susceptibility according to the EGMH (5). The contracture tests allowed to classify the patients as MHS (19), MHN (52). The others were classified as MH equivocal for halothane (7) and for caffeine (6).

RESULTS

Whereas, at rest, no differences were observed among patients and normal subjects, MH and EH patients developped an abnormal P31 spectral configuration during exercise and recovery.

1) For all subjects, ATP remained at near resting concentrations, even under ischemic exercise.

2) No differences appeared between control and MHN subjects.

TABLE 1 Changes in creatine phosphate (CP) concentration and intracellular pH (pHi) at rest, at the end of exercise (6') and during recovery (3' and 8'). Values are presented as mean ± standard deviation. Statistical significance was evaluated by Student's t test. *** $P < 0.001$; ** $P < 0.01$; * $P < 0.05$.

		Controls		MHN	
		E	IE	E	IE
Rest	CP	0.80±0.01	0.80±0.03	0.83±0.05	0.80±0.04
	pHi	7.04±0.02	7.04±0.01	7.05±0.04	7.02±0.07
Ex 6'	CP	0.60±0.05	0.48±0.04	0.57±0.06	0.50±0.05
	pHi	6.72±0.06	6.67±0.04	6.75±0.10	6.68±0.11
Rec 3'	CP	0.72±0.04	0.77±0.03	0.77±0.05	0.82±0.04
	pHi	6.87±0.11	6.82±0.04	6.90±0.07	6.79±0.09
8'	CP	0.77±0.02	0.80±0.01	0.83±0.03	0.80±0.05
	pHi	6.91±0.05	6.94±0.05	6.93±0.05	7.00±0.05

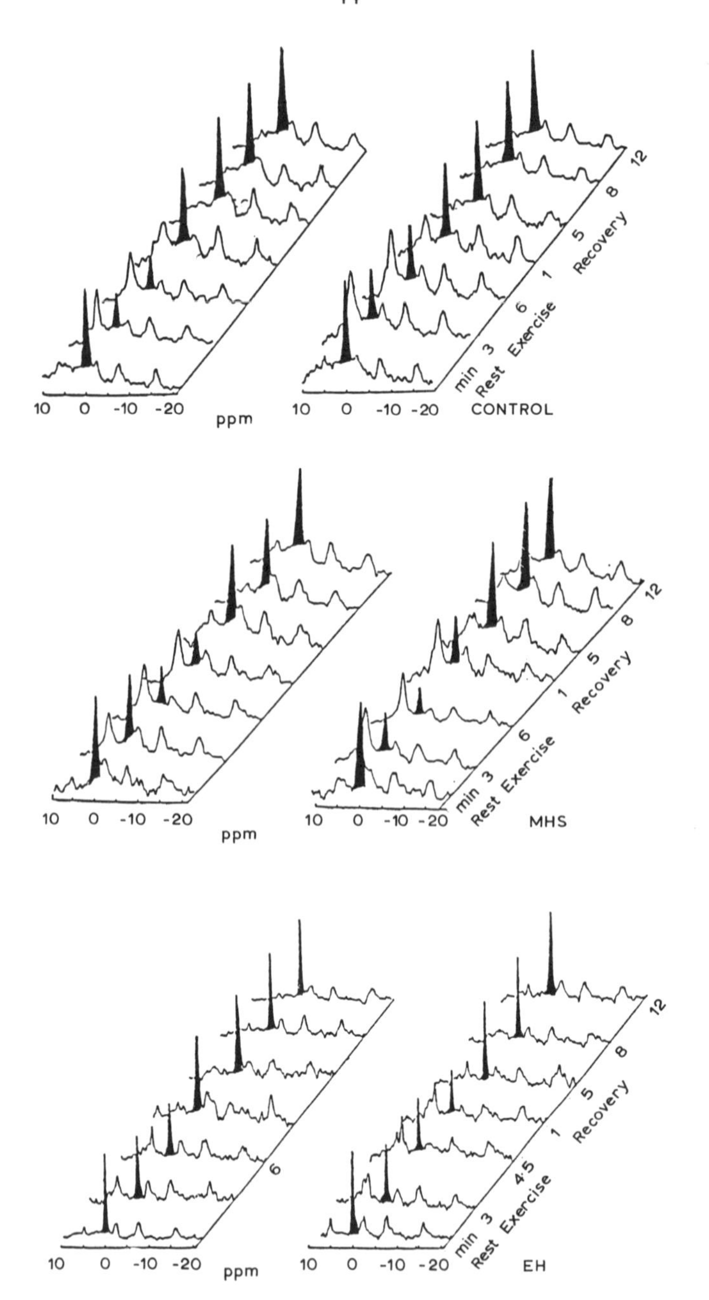

Fig. 2 Sequence of P31 NMR-S spectra from forearm muscles in a control subjects (control in a Malignant Hyperthermia Susceptible (MHS) patient and a patient who suffered for exercise hyperthermia (EH)).
CP peak is set at 0 ppM ; EH subject was unable to maintain exercise in hypoxia and had to stop at min 4.30 (*) ; A - spectra recorded in free arterial perfusion ; B - spectra recorded in restricted arterial perfusion (hypoxia) limited to the exercise period.

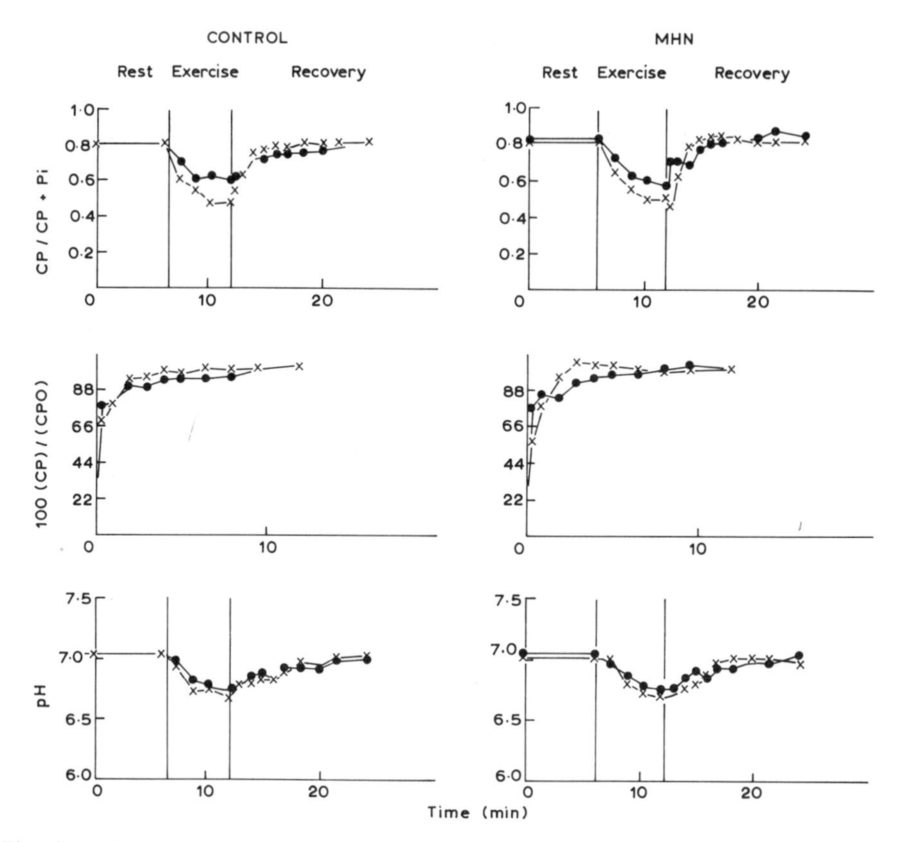

Fig. 3 Changes in pHi and concentration of creatine phosphate (CP) during exercise and recovery periods. □ : exercise in free arterial perfusion ; ▼ : exercise in restricted arterial perfusion (ischemia).

Figure 3 shows the kinetics of CP concentration, CP regeneration and pHi. CP concentration fell during the exercise period. The drop was larger in ischemic exercise, but limited to the half of the resting level. In control and MHN subjects, recovery to the resting metabolic state from exercise is rapid. Following an ischemic exercise, CP recovery is faster in the first 3 minutes. No significant differences were observed in the evolution of pHi during and after an exercise in normal or restricted circulatory conditions.

3) In MHS as in MH equivocal subjects, CP breakdown during a non ischemic exercise differed from those of control and MHN subjects.

TABLE 2 Changes in creatine phosphate (CP) concentration and intracellular pH (pHi) at rest, at the end of exercise (6') and during recovery (3' and 8').

		MHS		EH	
Rest	CP	0.82±0.04	0.82±0.03	0.81±0.02	0.80±0.01
	pHi	7.00±0.05	6.98±0.06	7.05±0.02	7.02±0.01
Ex 6'	CP	0.51±0.10	0.42±0.12	0.52±0.03	0.41±0.02
	pHi	6.47±0.09	6.23±0.07	6.36±0.04	6.13±0.04
Rec 3'	CP	0.77±0.06	0.75±0.09	0.74±0.01	0.70±0.02
	pHi	6.55±0.08	6.40±0.08	6.72±0.11	6.33±0.03
8'	CP	0.81±0.05	0.78±0.08	0.82±0.02	0.77±0.01
	pHi	6.79±0.07	6.77±0.06	6.94±0.05	6.70±0.07

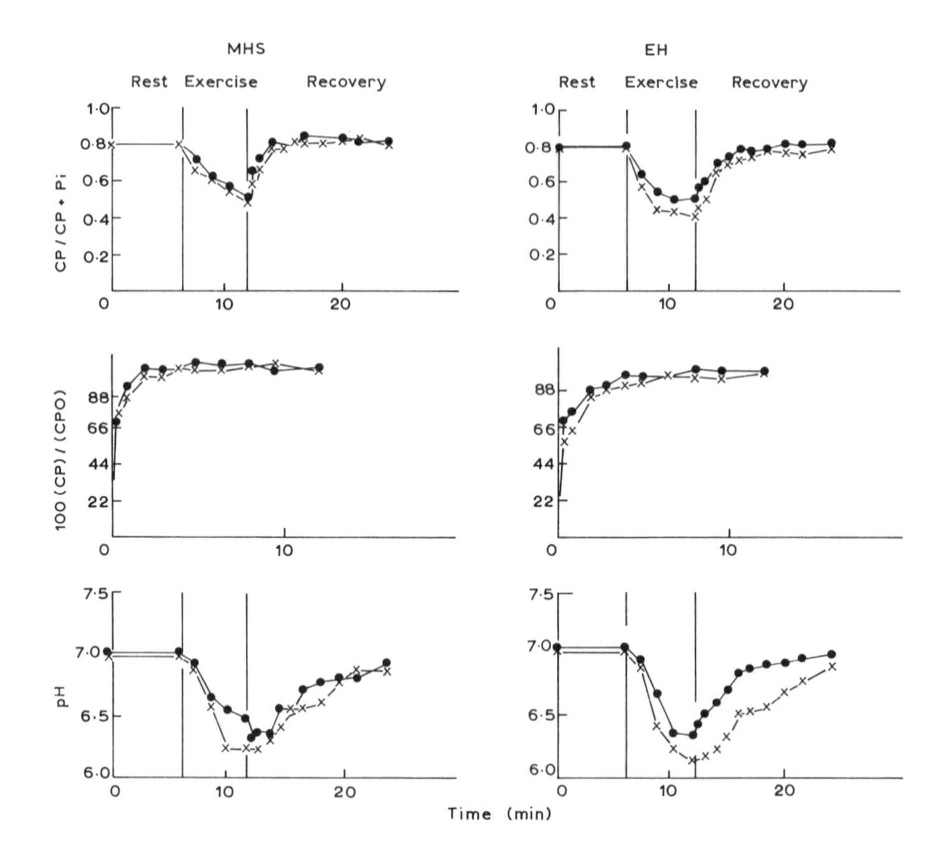

Fig. 4 Changes in pHi and concentration of creatine phosphate (CP) during exercise and recovery periods. □ : exercise in free arterial perfusion ; ▼ : exercise in restricted arterial perfusion (ischemia).

Moreover, an ischemic exercise produced a faster and larger drop in the first 3 minutes. But a plateau could be attained in the last 3 minutes. At the onset of recovery, CP resynthesis was slower and ischemia did not modify the rate of resynthesis.

Striking differences were observed in pHi. During an ischemic or not exercise, a fast and large drop in pHi was observed. The return to rest values was slowed and at the end of the recovery period a persistant acidosis was observed (pHi : 6.7 g/6.77).
4) All Exercise Hyperthermia subjects presented the same abnormalities. With normal blood flow, during an exercise the drop in pHi was increased, but the return to rest value was suitable. Discrepancies between control and EH subjects were observed during and after an ischemic exercise. Ischemia did not accelerate the CP resynthesis : the pHi evolution was similar to those of MHS subjects : a large drop and a lengthened recovery (Table 2).

DISCUSSION

The results indicate a perturbed energy metabolism in EH and MHS subjects. The presence of metabolic alterations is not very surprising in MH since its suspected mechanism is an excess release of calcium ions which are known to activate glycolysis simultaneously with muscle fiber contraction. Contraction can increase ATP hydrolysis instantaneously by many-fold and the cell must be able to respond quickly to this demand for energy. Transfer of high energy phosphate from CP to ATP via the creatine kinase reaction is the immediate source to maintain ATP level. At the onset of a contraction, CP concentration fells. Glycogenolysis and glycolysis are switched on by Ca2+ when contraction begins. This anaerobic source of ATP is not sufficient to maintain the CP concentration. It is only by oxydative phosphorylation that a sufficient amount of ATP is produced. At this step, the CP level is in a "plateau phase" until the contraction ceases. At the onset of the recovery, the CP supply is replenished by net rephosphorylation of creatine. If the exercise performed is exhaustive or if one of these energetic pathways are perturbed, CP concentration fells until the end of contraction.

Some muscle disorders are known to result from metabolic blocks in glycolytic energy pathway (7) and in some of the mitochondrial myopathies (3). In others, such as Duchenne muscular dystrophy, alterations in metabolite concentrations have been reported (14). Comparison of results from normal subjects and from patients with defects in oxidative pathway can be expected to provide information on the control of energy production and utilisation in skeletal muscle. In patients with congestive heart failure (11), during an exercise the CP is depleted and the pHi low (6.25). In MH and EH subjects, particularly during an ischemic exercise, a large drop in CP concentration was observed and the "plateau phase" appeared later. Simultaneously, the pHi fall indicate an active anaerobic metabolism.

Metabolic disorders observed in MH and EH subjects can indicate defects in oxydative metabolic pathways leading to a prevalent glycolytic pathway.

CONCLUSION

P31-SRM is the method able to detect non invasively abnormalities of muscular energetic metabolism linked or not to a known myopathy. Even in the absence of crisis, P31-SRM reveal a latent myopathy in MHS as in EH subjects.

P31 NMRS is a useful non invasive tool both from the standpoint of ability to study and to detect diseased muscles, and from the standpoint of patient's safety because the measurements, unlike biopsies, are harmless.

REFERENCES

1. Barany, M., Doyle, D.D., Graff, G., Westler, W.M. and Markely, J.L. 1984. Natural abundance of 13C NMR spectra of human muscle, normal and diseased. Magnetic Resonance in Medicine, 1, 30-43.
2. Barany, M. and Glonek, T. 1983. Identification of diseased states by phosphorus-31 NMR. In : Phosphorus-31 NMR. Principles and applications. (Ed. D.G. Gorenstein) Academic Press. pp. 511-544.
3. Chance, B., Leigh, J.S., Smith, D.S., Nioka, S. and Clark, B.J. 1986. Phosphorus magnetic resonance spectroscopy studies of the role of mitochondria in the disease process. In : Annals of the New York Academy of Sciences, 488, 140-153.
4. Chance, B., Younkin, D.P., Kelley, R., Bank, W.J., Berkowitz, H.D., Argov, Z., Donlon, E., Boden, B., McCully, K., Buist, N.M.R. and Kennaway, N. 1986. Am. J. Med. Gen., 25, 659-679.
5. The European Malignant Hyperpyrexia Group, Ellis, F.R. et al. 1984. A protocol for the investigation of Malignant Hyperpyrexia (MH) susceptibility. Br. J. Anaesth., 56, 1267.
6. Goldman, M. and Servoz-Gavin, P. 1986. La résonance magnétique nucléaire. Clefs CEA, n°1.
7. Jehenson, P., Duboc, D., Fardeau, M. et Syrota A. 1987. Etude in vivo par spectroscopie RMN du phosphore 31 du métabolisme musculaire en pathologie et sous l'influence de thérapeutiques. Thérapie, 42, 467-470.
8. Kozak-Reiss, G., Desmoulin, F., Canioni, P., Cozzone, P., Gascard, J.P., Monin, G., Pusel, J.M., Renou, J.P. and Talmant, A. 1987. In : Evaluation and control of meat quality in pigs. (Eds Tarrant P.V., Eikelenboom G., Monin G.). Martinus Nijhoff. pp. 27-38.
9. Kozak-Reiss, G., Gascard, J.P. et Redouane-Bénichou, K. 1986. Dépistage de l'Hyperthermie Maligne anesthésique par les tests de contracture musculaire et par la spectroscopie RMN (SRM-31P). Ann. Fr. Anesth. Réanim., 5, 584-589.
10. Kozak-Reiss, G., Gascard, J.P., Hervé, Ph., Jehenson P. and Syrota, A. 1988. Malignant and exercise hyperthermia investigation of seventy subjects by contracture tests and P31 NMR spectroscopy. Symposium sur l'Hyperthermie Maligne. Vienne (Autriche), 8-13 mars 1988.
11. Massie, B.M., Conway, M., Yonge, R., Frostick, S., Sleight, P., Ledingham, J., Radda, G. and Rajagopalan, B. 1987. 31P nuclear magnetic resonance evidence of abnormal skeletal muscle metabolism in patients with congestive heart failure. Am. J. Cardiol., 60, 309-315.
12. Olgin, J., Argov, Z., Rosenberg, H. and Chance, B. 1987. Non-invasive detection of malignant hyperthermia susceptibility by in vivo 31P NMR. Anesthesiology, 67, A166.
13. Radda, G.K. and Taylor, D.J. 1985. Applications of nuclear magnetic resonance spectroscopy in pathology. In : International review of experimental pathology. (Eds Richter C.W., Epstein M.A.). Academic Press. vol. 27. pp 1-58.

14. Younkin, D.P., Berman, P., Sladky, J., Chee, C., Bank, W. and Chance, B. 1987. 31P NMR studies in Duchenne muscular dystrophy : age-related metabolic changes. Neurology, 37, 165-169.

DISCUSSION

Chairperson: A. Ganssen/FRG

Convincing examples of a variety of MR-imaging applications in medical diagnostics in man were presented by P. Heintz (FRG). Especially the unique soft tissue contrast combined with high spatial resolution and the possibility of arbitrary imaging plane selection give new insights into the live human body never experienced before. The importance of the proper choice of the signal acquisition parameters for the reproduction of lesions was demonstrated. While the sensitivity with respect to the imaging of small changes in tissue properties can be extremely high, the specificity for malignancy is often questionable. The disadvantage of the comparatively long data acquisition times giving rise to motion unsharpness in the case of moving organs can be partially overcome in the cases of periodic motion like for the imaging of the heart or the thorax by ECG- or respiratory gating methods. While NMR imaging can already be considered as the method of choice for most neural and spinal diagnostic problems. In its application to the great variety of diagnostic problems it has to be compared with the present potential of the other diagnostic imaging modalities - especially with regard to cost efficiency.

Well resolved localized proton spectra of the human brain and muscle - in vivo - were demonstrated by J. Frahm/FRG. Volumes of interest of only a few cm^3 could be investigated within minutes at a field of 2.35 T with the STEAM-sequence, and simultaneous water suppression by the additional application of chemical shift selective r.f. pulses. The most important metabolites like lactate, N-acetylaspartate, phosphocreatine, and amino acids, fatty acids, lipids showed up in natural abundance, thus enabling non-invasive, in vivo - metabolic studies, and possibly the metabolic assessment of malignancies.

Malignant hyperthermia is a serious congenital syndrome in pigs. The energy metabolism involved including intracellular pH-value can be followed-up in vivo and non-invasively by P 31-NMR-spectroscopy as has been reported by G. Kozak-Reiss/France. Malignant hyperthermia and excercise hyperthermia have been investigated in a comparative study in man and in pigs. The difference between normal and malignant hyperthermia subjects has been a faster and deeper decrease in pH and a slower return to the resting value after exercise. A perturbed energy metabolism with an increased anaerobic glycolysis has been demonstrated with the P 31-spectra in malignant hyperthermia as well as in exercise hyperthermia. Metabolic abnormalities have also been observed by extended recovery periods after ischemic exercise.

SESSION III

BODY FLUIDS AND CIRCULATION

Chairperson: P. Heintz

NMR Spectroscopy of Body Fluids in Medical Diagnosis

Werner Offermann and Dieter Leibfritz
Fachbereich Chemie/Biologie, Universität, Bremen, West Germany

Contents:

Introduction

Medical diagnosis and nuclear magnetic resonance span the entire range of the scientific scale. Medical diagnosis deals with human existance, NMR is nuclear physics and high tech. In order to avoid the specialized idioms of both, let us consider the relation between them in basic scientific terms.

Disease, which is the main object of medical diagnosis, may be defined as a deviation from the 'normal' steady state thermodynamics[1], which characterize a given (healthy) biological system[2]. In turn, an NMR spectrum reflects, at least partially, the thermodynamics of the sample as will be shown below.

This admittedly strange and abstract biophysical view is very helpful in order to understand which kind of information NMR may contribute to the life sciences. Since a single NMR spectrum contains, as a rule, no kinetic information, the regulatory mechanisms which maintain the 'normal' steady state of a biological system can only be investigated by more sophisticated NMR techniques such as spin labelling[3]. Also, the impact of 'soft' stimuli on biological systems, such as the application of drugs in physiological doses, mild hypo- or hyperthermia, and the internal trigger of the heart beat remains invisible in simple NMR spectra as it is immediately counterbalanced towards homeostasis. Again, special time consuming methods, *e. g.* the double stimulus experiment[4], must be employed. The potential value of NMR spectroscopy in medical diagnosis lies in the fact, that, owing to the relation formulated above, the most simple NMR experiment suffices to deliver the required information.

^{31}P NMR Spectroscopic Study of Blood Conservation

The results used in this section were obtained by W.B. WERK and P.B. VIDYASAGAR and presented during the XIII ICMRBS[5].

Figure 1 shows the ^{31}P NMR spectrum of human blood. It serves here only to demonstrate the information content of a simple spectrum. The signals indicate the

---

[1] L. v. BERTALLANFFY: Theoretische Biologie, pp 58. Francke, Berlin 1951.

[2] V. BECKER, K. GOERTTLER, and H.H. JENSEN: Konzepte der theoretischen Pathologie, pp 25. Springer, Berlin 1980. W.W. HÖPKER: Das Problem der Diagnose und ihre operationale Darstellung in der Medizin, p 3. Springer, Berlin 1977.

[3] The available methods are briefly described in: P. RÖSCH: NMR Studies of Phosphoryl Transferring Enzymes. *Progr. NMR Spectr.* **18** 123-169 (1986).

[4] W. OFFERMANN, W. KUHN, S. SOBOLL, T. ISHIKAWA, and D. LEIBFRITZ: The *in Vivo* Contour Plot. An Improved Representation of Stimulus Experiments. *Magn. Reson. Med.* **4** 507-516 (1987).

[5] W.B. WERK, P.B. VIDYASAGAR, and D. LEIBFRITZ: A ^{31}P-NMR Study of the Preservation of Blood. *Proc. XIII Int. Conf. Magn. Reson. Biol. Syst.* P4-8 (1988).

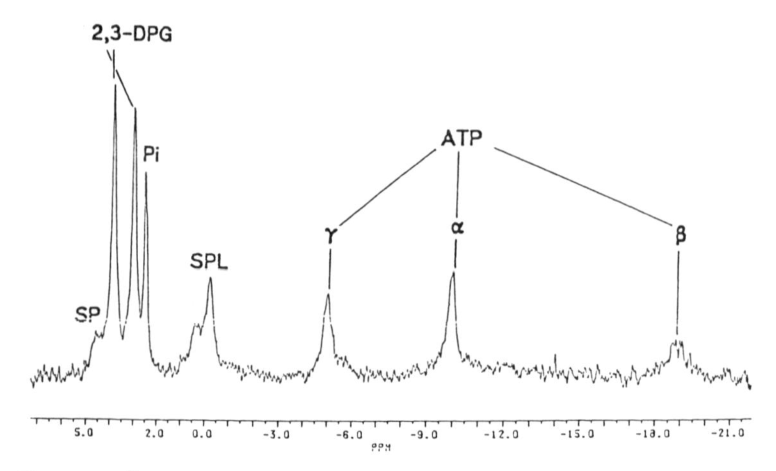

Figure 1. ^{31}P NMR Spectrum of Human Blood. The Signals Indicate the Presence of the Following Components: SP, Sugar Phosphates; 2,3-DPG, 2,3-Diphosphoglycerate; P_i, Inorganic Phosphate; SPL, Serum Phospholipids; ATP, Adenosinetriphosphate. The Signals of γ- and α-ATP Coincide with those of β- and α-ADP. Signal Areas are Proportional to Concentrations.

presence of various compounds containing phosphate residues. If the spectrum was acquired properly, the area underneath a signal is proportional to the number of resonating nuclei, *i. e.* to the concentration of the respective compound. Hence, integration yields the relative concentrations which already suffice to calculate, *e. g.*, the equilibrium constant of the dephosphorylation of ATP. Absolute concentrations are obtained by comparison to a reference compound. From the position of the P_i signal the *p*H can be deduced[6]. In this study, the gradual shift of the concentration of phosphorylated compounds away from the 'normal' values was used as a quality control.

Identification of Neurological Diseases by ^{1}H NMR of Cerebrospinal Fluid

The results in this section were obtained in cooperation with F. KOSCHOREK, H. GREMMEL (both Radiologische Universitätsklinik, Kiel, FRG), E. KRÜGER (Neurochirurgische Universitätsklinik, Kiel, FRG), W. WOSNIOK, F. LIERMANN, J. TIMM (all Fachbereich Mathematik/Informatik, Universität, Bremen, FRG), T. REESE, and J. STELTEN (both Fachbereich Chemie/Biologie, Universität, Bremen, FRG), and have been published in part in *Radiology*[7].

Cerebrospinal fluid (CSF) is taken routinely from the lumbar spine of patients suspected of neurological disease. We analyzed the ^{1}H NMR visible contents of CSF mainly from patients with subcutaneous bleeding, disc herniation, and brain tumors in order to find 'marker' metabolites which are typical for these diseases and might be helpful in neurological diagnosis. We quantified 16 metabolites and could reassign by preliminary discrimination analyses 85 % or more of the cases, which is by no means sufficient but at least promising.

Experimental. For ^{1}H NMR spectroscopy, 1 ml of CSF was lyophilized and resuspended in 0.5 ml D_2O. Spectra were acquired in less than ten minutes at 360

---

[6] R.A. ILES, A.N. STEVENS, and D. SHAW: NMR Studies of Metabolites in Living Tissue. *Progr. NMR Spectr.* **15** 49-200 (1982).
[7] F. KOSCHOREK, H. GREMMEL, J. STELTEN, W. OFFERMANN, and D. LEIBFRITZ: Cerebrospinal Fluid: Detection of Tumors and Disk Herniations with MR Spectroscopy. *Radiology* **167** 813-816 (1988).

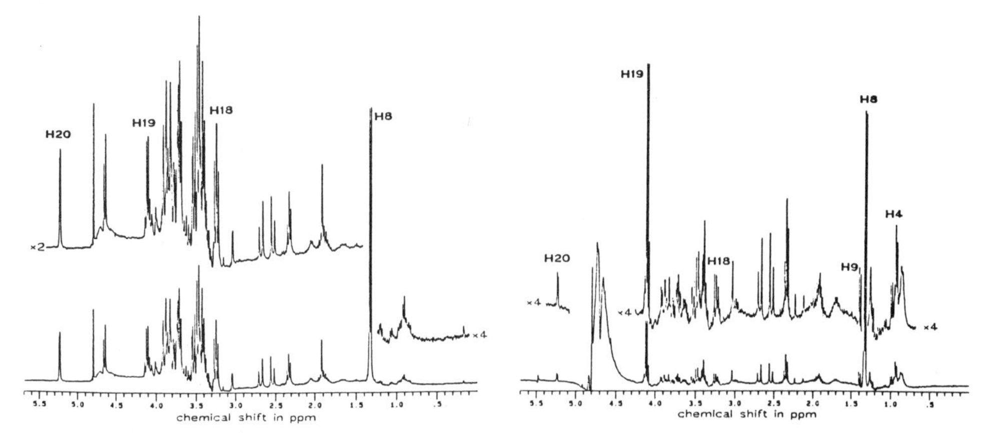

Figure 2. ^{1}H NMR Spectrum of Human CSF. Left, Normal Sample, Right, Tumorous Sample. Component Levels were Normalized to Lactate (Signals H8 and H19). In the Tumorous Sample, Glucose (H18 and H20) is Lowered, α-Alanine (H9) and Valine (H4) are Increased.

MHz on a Bruker AM 360 WB or a Bruker WH 360, spectrometers, which are available in most chemistry departments. The still large residual HDO signal was suppressed by selective presaturation. Two of the resulting spectra are depicted in Figure 2. The appearance of ^{1}H NMR spectra is complicated by a hyperfine splitting due to *J*-coupling with the consequence that the signal assigned to a biochemical species will, as a rule, consist of several lines. When this splitting and also the number of resonating nuclei per molecule is properly allowed for, relative concentrations are obtained. We normalized all the data by setting the lactate (signal H8) in each sample to 100 per cent.

Normal profile. The data include 20 cases, where the initial suspicion of disease could not be confirmed. We regard these data as representative for healthy CSF.

Table 1. The Normal Profile of Components in Human CSF which are Visible to ^{1}H NMR Spectroscopy.

Signal*	Chemical Shift (ppm)†	Assignment‡	Relative Concentration§
H1	0.17	CH_3, not assigned	3.05 ± 1.75 (13)
H2	0.85	CH_3, lipids	1.70 ± 1.50 (19)
H4	0.98, 1.03	CH_3, valine	0.31 ± 0.47‖ (18)
H5	1.18	CH_3, 3-hydroxy butyric acid	2.25 ± 1.59 (19)
H6	1.26	CH_2, phospholipids	1.25 ± 1.89‖ (18)
H8	1.33	CH_3, lactate	100.00 ± 0.00 (19)
H9	1.45	CH_3, α-alanine	1.76 ± 1.23 (19)
H10	1.92	CH_3, acetate	7.59 ± 4.05 (19)
H11	1.97	CH_2?, not assigned#	2.60 ± 1.40 (18)
H12	2.02	CH_2?, not assigned	1.20 ± 0.70 (18)
H13	2.24	CH_3, acetone	...
H14	2.39	CH_2, not assigned#	14.34 ± 5.04 (19)
H15	2.62	CH_2, citrate	10.61 ± 2.09 (19)
H16	3.04	CH_3, creatinine?	3.62 ± 0.58 (19)
H17	3.05	CH_3, creatine?	2.80 ± 1.08 (19)
H18	3.25	C_2-H, β-D-glucose	190.80 ± 26.60** (19)
H20	5.24	C_1-H, α-D-glucose	
H19	4.12	CH, lactate	...

* The signals are numbered arbitrarily. The numbers are the same as in Figures 1 and 2.
† The signal positions are referenced to external tetramethyl-silane (0.0 ppm) by setting the methyl doublet of lactate (H8) to 1.33 ppm.
‡ Preliminary assignments are indicated by a question mark.
§ Given as weighted mean ± standard deviation, with the number of samples in parentheses. Lactate was set arbitrarily at 100.
‖ Unweighted mean.
\# Not glutamic acid.
** Sum of both glucose anomers.

Table 1 is a (still incomplete) list of the CSF components we found, together with their relative concentrations.

Diseases. A superficial comparison of two randomly chosen samples of CSF, one healthy, one tumorous (Figure 2), suggests that some component levels are different and others are not. The tumorous sample is definitely lower in glucose (signals H18 and H20), but higher in the amino acids alanine (H9) and valine (H4). In Figure 3, the glucose levels of all 'normal' and tumorous samples are compared. Indeed, the average glucose content, or, more precisely, the average ratio of glucose to lactate is lowered in the CSF of tumor patients. This observation may indicate enhanced energy metabolism *via* anaerobic glycolysis caused by the tumor.

Discrimination analyses. Apparently, the overlap between normal and tumorous glucose levels as depicted in Figure 3 is too large for diagnostic purposes. In order to include all data into the analysis and to eliminate the less relevant components, we tried various methods of discrimination analysis. These procedures try to locate the locus of each patient group within the sixteen-dimensional parameter space spanned by all component levels. The quality of the search is then checked by reassigning all data sets (of course without the clinical diagnosis) to patient groups. A linear method yielded sufficient results for three groups of cases but found only 61 % of the tumors (Table 2). A nearest neighbor method did better with the tumors but failed with the other groups.

Presently, we are expanding our study to a greater number of case groups and patients. In order to improve the statistical analysis, we include more parameters such as the age of the patients, and we preprocess the data so that they meet the requirements of discrimination analyses. We are confident, that finally ^{1}H NMR will aid neurological diagnosis as well as the pathophysiology of these diseases.

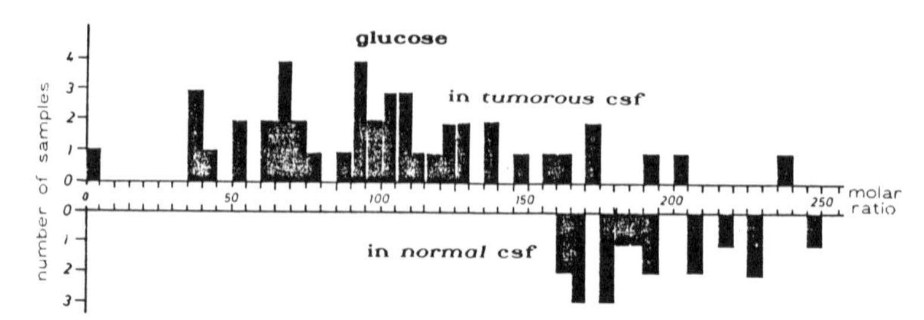

Figure 3. Glucose Levels in Normal and Tumorous CSF. The Average of the Tumorous Samples is Lower, but Both Populations Overlap Sufficiently.

Table 2. Discrimination analysis of 143 samples of human CSF characterized by 16 component levels determined by ^{1}H NMR. The bold face values indicate the quality of the reassignment. The line totals are not identical as the nearest neighbor method did not classify some samples at all.

From Diagnosis	Assigned to Bleeding	Disc	Tumor	Normal
		Linear Method		
Bleeding	**48 (98%)**		1 (2%)	
Disc		**23 (100%)**		
Tumor	20 (39%)		**31 (61%)**	
Normal	2 (10%)	1 (5%)		**17 (85%)**
		Nearest Neighbor Method		
Bleeding	**30 (61%)**		15 (31%)	
Disc		**8 (35%)**	4 (17%)	2 (9%)
Tumor	2 (4%)		**44 (86%)**	3 (6%)
Normal	1 (5%)	4 (20%)	5 (25%)	**7 (35%)**

The Fossel Test - No Fast Screening for Tumors

The results in this section were obtained in cooperation with O. SEZER, H. RASCHE (Klinik f. innere Medizin, Zentralkrankenhaus, Bremen, FRG), D. SEIDEL (Zentrum f. innere Medizin, Universität, Göttingen, FRG), and J. STELTEN (Fachbereich Chemie/Biologie, Universität, Bremen, FRG).

The rise and fall of the Fossel test is one of the comparatively rare stories where human emotion is closely interwoven into science. Therefore, we present it as a story rather than as a scientific report.

Immediately after its publication[8], the Fossel test attracted the attention not only of many scientists and physicians but also of the general public, because of its potential as a fast screening method for malignant tumors. The test required an ^1H NMR spectrum of blood plasma recorded in the way we described for CSF above. Plasma spectra are superficially similar to CSF spectra, but Fossel and his co-workers restricted their interest to the two broad signals near 1.25 and 0.85 ppm, which arise from the methylene and methyl groups of the various lipoprotein fractions in blood (these signals appear also in the spectrum of tumorous CSF in Figure 2). They determined the average width of these two lines, measured at half signal height, and claimed that patients with malignant tumors could be recognized by a low value around 30 Hz, whereas persons with non-tumor and without any diseases had mean line widths around 36 and 40 Hz.

The groups of C. E. Mountford and I. C. P. Smith had investigated the lipoprotein signals with respect to the physical state of the donors before (and remained unquoted in reference[8]), but Fossel *et al.* were the first to define the threshold at 33 Hz, which separates tumorous from non-tumorous cases with high specificity. On a spectrometer equipped with an automatic sample changer, approximately 100 to 150 samples might be measured per day, and the diagnosis be made by a simple ruler: what prospects in medical care! what a market for spectrometers!

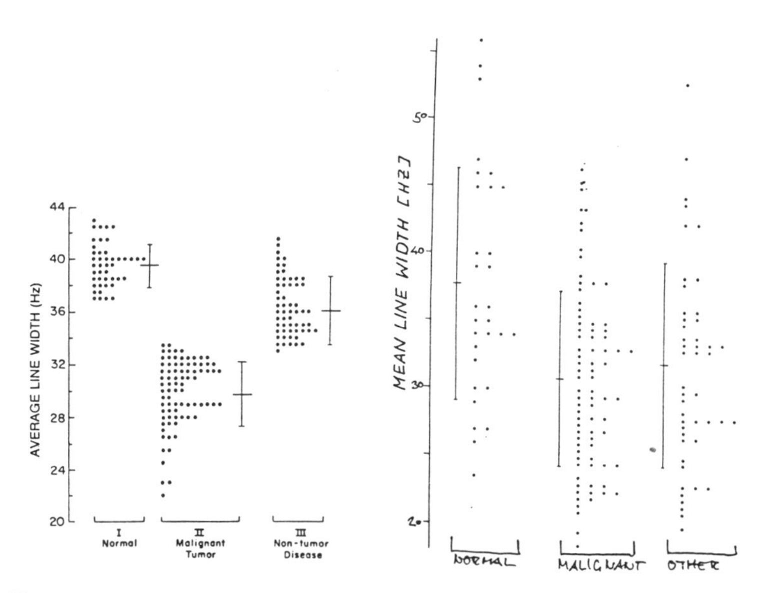

Figure 4. Distribution of the Fossel Index (Mean Lipid Line Width) in Three Different Populations as Observed by Fossel[8] (Left) and by Us[10] (Right). Note the Following Differences: The Indices of the 'Other' Diseases, the Width of the Distributions, and the Shape of the Distributions.

However, other groups could not reproduce these results. Healthy and tumorous line widths were not so well separated, and non-tumorous diseases fell in between. In Figure 4, Fossel's data[8] (left) are compared to ours as an example. So, looking at a messy pattern of data and not being prepared to regard a scientific report as faked, most of the involved researchers cast their main interest on the experimental protocol to be followed in order to reproduce Fossel's results. The controversy culminated in a special session devoted to the Fossel test during the sixth annual meeting of the Society of Magnetic Resonance in Medicine, held in August 1987 in New York.

Fossel defended his index with data collected from more than 1000 patients. They were distributed like his previous results in Figure 4, with a slightly increased standard deviation. G. N. Chmurny[9], D. Leibfritz[10], and T. R. Brown[11] reported to have found similar line widths in malignant and non-tumorous samples and a larger standard deviation in all groups leading to sufficient overlap between healthy and tumorous samples. Reassignment arrived at numerous false positive results (*i. e.* non-tumorous samples were identified as tumorous), mainly from the other diseases but also from certain dispositions such as pregnancy and smoking. The asymetric distribution curves in Fossel's groups were challenged, because the data pile up so conveniently just short of the separation line, even when more than 1000 samples are regarded. The significance of the line width parameter itself was questioned, as it is measured on a 'line' which is actually the sum of several overlapping lines having different positions, widths, shapes, and intensities.

Of the sixteen contributions, which referred to various aspects of the Fossel test, only two (except Fossel's own) supported the validity of the data, while seven doubted it. Several expansions such as the inclusion of line intensities, ^{13}C NMR data, T_2 measurements, and chromatographic parameters were suggested to improve the specificity of the method. However, the original beauty of the fast and simple procedure had vanished, together with the hope of a general tumor screening.

---

[8] E.T. FOSSEL, J.M. CARR, and J. McDONAGH: Detection of Malignant Tumors. *N. Engl. J. Med.* **315** 1369-1376 (1986).

[9] G.N. CHMURNY, M. MELLINI, D. HALVERSON, N. CAPORASO, W. URBA, G. MUSCHIK, H.J. ISSAQ, B.D. HILTON, I.C.P. SMITH, M. MONCK, S. ADAMS, T. KROFT, M. DOVER, M. PRÉFONTAINE, and J.K. SAUNDERS: Simple Blood Tests for Human Cancer: Evaluation of NMR and Other Analytical Methods. *Sixth Ann. Meeting. Soc. Magn. Reson. Med.* W28 (1987).

[10] O. SEZER, J. STELTEN, W. OFFERMANN, H. RASCHE, D. SEIDEL, and D. LEIBFRITZ: Line Width of Proton NMR Spectra of Human Plasma: An Efficient Method to Detect Cancer? *Sixth Ann. Meeting Soc. Magn. Reson. Med.* W29 (1987).

[11] S.D. BUCHTHAL, M.A. HARDY, and T.R. BROWN: Assessing the Value of Screening for Malignant Disease by Proton NMR Spectroscopy of Plasma. *Sixth Ann. Meeting Soc. Magn. Reson.* W30 (1987).

NMR FLOW IMAGING - AN OVERVIEW

David G Norris
Universität Bremen, West Germany.

INTRODUCTION

This article describes the three major effects that flow can have upon the NMR imaging experiment, and their application to methods of flow-imaging. The images presented here were obtained using the Aberdeen Mark1 system (1,2), which operates at the extremely low field strength of 0.04T. As a consequence the images suffer from a poor signal to noise ratio, and hence do not represent the state of the art in NMR flow imaging, but merely serve as illustrative examples.

MOTION INDUCED PHASE EFFECTS

In discussing phase effects due to motion, an important concept is that of the 'isochromatic spin group' first developed by Hahn (3). An isochromatic spin group or isochromat, is a group of nuclei that always experience the same local magnetic field. The number of nuclei within be the group is considered to be large enough for its net magnetization to treated classically. Phase changes in the magnetization of such a group will occur if it moves in the direction of a magnetic field gradient. In such a situation the phase will change at a rate given by

$$\frac{\partial}{\partial t}\varphi(t) = \gamma \mathbf{r}.\mathbf{G} \qquad [1]$$

where $\varphi(t)$ is the phase of the isochromat at time t, **r** its position vector, and **G** the magnetic field gradient.

In phase-encoded flow imaging a magnetic field gradient is applied in such a way that the phase of stationary isochromats is unaltered, whilst those moving with a constant velocity undergo a phase change directly proportional to that velocity. In order to determine the conditions that such a gradient must satisfy, consider a 1-dimensional situation, in which an isochromat moves at a constant velocity v, in the presence of a magnetic field gradient G_x. Its position at time t is then given by

$$x(t) = x_0 + vt \qquad [2]$$

x_o being the position at time t=0. By integration of equation 1 the phase at time T is

$$\varphi(T) = \int_o^T \gamma(x_o + vt)G_x(t)dt \qquad [3]$$

For stationary isochromats to experience no net phase change then

$$\gamma x_o \int_o^T G_x(t)dt = 0 \qquad [4]$$

and in this case the phase change for moving isochromats is

$$\varphi(T) = \gamma v \int_o^T tG_x(t)dt \qquad [5]$$

This equation shows that the velocity-induced phase change is proportional to the first moment of the gradient with respect to time.

A velocity-encoding pulse must satisfy equation 4. The simplest gradient that achieves this is a symmetrical bipolar pulse shown in figure 1a. This is equivalent to two identical monopolar pulses separated by a 180° rf pulse, shown in figure 1b. For a field gradient having this form equation 2.5 becomes

$$\varphi(T) = -\gamma v G_x t_A t_B \qquad [6]$$

i.e. the phase is directly proportional to the velocity.

The first coherent scheme for exploiting this effect in an imaging sequence was proposed by Moran (4). The basis of this method is either to incorporate velocity-encoding pulses in an imaging sequence, or to utilize the velocity-encoding nature of some of the imaging gradients. In the simplest form of this technique the phase of the image, rather than its magnitude is displayed. The phase, as is indicated in equation 6, is proportional to the velocity component in the direction of the velocity-encoding pulse.

This technique has the advantage that velocity-encoding pulses may be applied along any gradient direction. By applying velocity-encoding pulses along each of three orthogonal axes in turn, and taking the

a

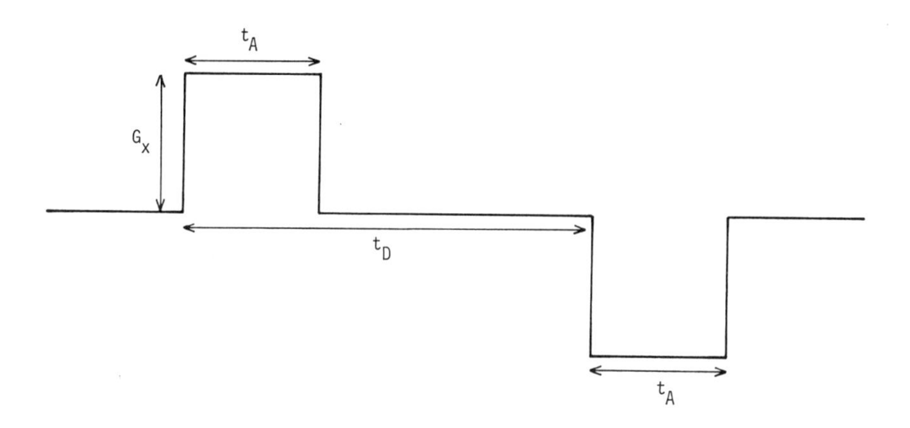

b

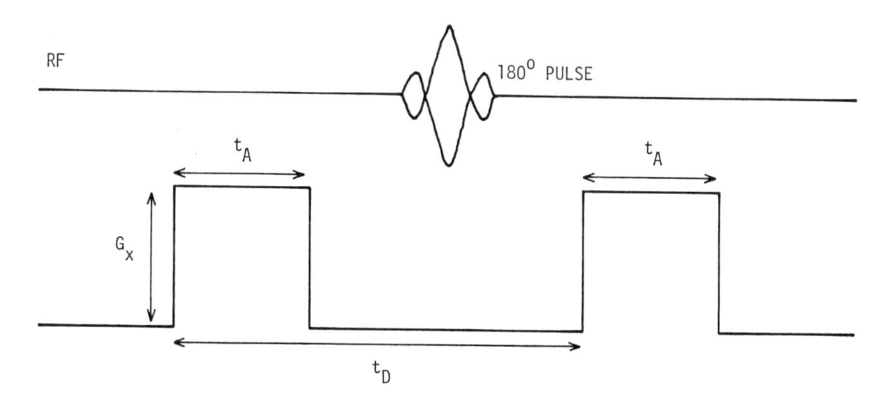

1. Velocity-encoding gradient pulses.

vector sum of the resulting velocity components, the absolute velocity may be determined. Or by conducting just one experiment the velocity component in any direction can be obtained.

A disadvantage is the possibility of velocity aliasing. This occurs when the phase change induced by the velocity-encoding gradient is greater than or equal to $+/-\pi$. There are two solutions to this problem: The first is to sample the phase with a sufficient frequency to satisfy the Nyquist theorem (5). The second is to assume that the velocity is zero at the vessel wall, and increases continuously as one progresses deeper into the vessel (6).

The phase of an NMR image may be dependent upon flow, main field inhomogeneity, and chemical shift. The effects of the last two of these are best eliminated if two image data sets are collected, having differing velocity sensitivities, but the same echo time (TE). The difference in phase between the two data sets should then be proportional only to the velocity.

Equation 3 showed how the motion of an isochromat in a gradient causes a change of phase. In the case of motion in the imaging gradients this may give rise to unwanted effects. The most significant of these being when the range of velocities across a voxel is sufficiently large that the corresponding range of encoded phases is of the order of, or greater than π. In this case the isochromats within the voxel will not be phase coherent, and there will be little or no net signal from that voxel (7). This problem may be resolved by flow-stabilizing the gradient in question (6). The principle of flow stabilization is to add extra gradient pulses so that there is no extra phase change for moving isochromats. This not only prevents signal loss, but eliminates unwanted phase changes due to motion in the imaging gradients.

The lowest velocity detectable by this technique depends upon the strength of the gradient that can be applied, and its maximum duration. The strength of the gradient is determined by the capabilities of the system, the maximum duration may not exceed the echo-time, and is thus heavily dependent upon the type of pulse sequence employed, and the T_2 or T_2^* of the object.

The upper velocity limit is dependent upon the direction of flow relative to the slice. For flow in the slice there is no upper limit, other than that at which the imaging process itself becomes unacceptably distorted. For flow through the slice, the normal limitation is due to loss of signal caused by isochromats leaving the slice within TE.

An example of a flow image obtained using the phase-encoding method is shown in figure 2 . The phantom consisted of a pair of plastic tubes, containing doped water flowing at 0.1m/s in opposite directions. Two images were obtained with bipolar field gradient pulses applied along the length of the tubing. The polarity of the pulses was reversed between the two images, and the phase difference obtained is shown in the figure. The flow within the tubes was laminar, and this is reflected by the greater phase difference at the centre than at the edges. The background is phase noise. This method is capable of measuring velocites to an accuracy of better than 5%, and has the highest intrinsic accuracy of all the NMR flow imaging methods.

It is possible to extend the phase-encoding technique to higher derivatives of veloctity. For example, if two bipolar pulses are placed back to back, as shown in figure 3, then the phase change induced in isochromats moving with constant velocity will be zero, as the phase change induced by the first pair of pulses will be exactly cancelled by that induced by the second pair. If an isochromat is moving with constant acceleration then its position at time t will be

$$x(t) = x_o + vt + at^2/2 \qquad [7]$$

where v is the velocity at t=0 and a the acceleration. The phase change for accelerating isochromats is then

$$\varphi(T) = (\gamma a/2)\int_0^T t^2 G_x(t)dt \qquad [8]$$

which for the pulse sequence shown in fig 2.2 gives a result of

$$\varphi(T) = 2\gamma a G_x t_A^3 \qquad [9]$$

Equation 5 showed that the velocity-induced phase change is proportional to the first moment of the gradient, equation 9 shows that the acceleration-induced change is proportional to the second moment.

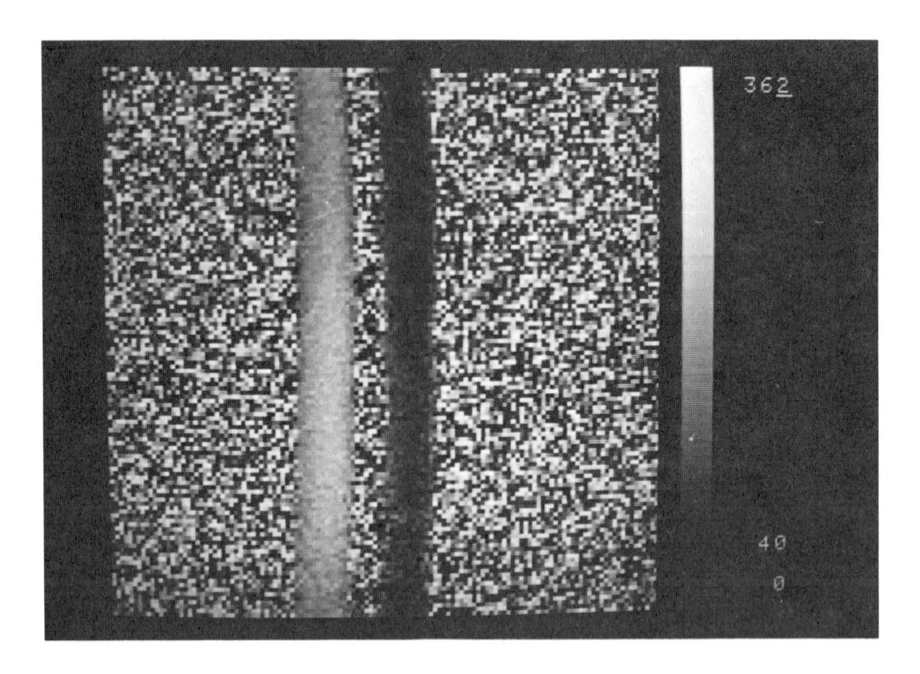

2. Phase difference images showing laminar flow in a phantom.

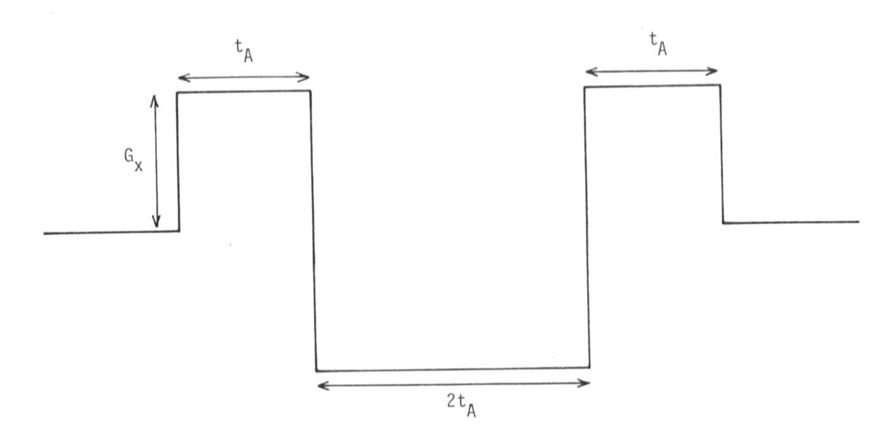

3. Acceleration-encoding gradient pulses.

This pattern extends to higher derivatives of position.

It is also possible to use balanced biolar pulses to phase-encode velocity, by a method directly analagous to that in which one spatial coordinate is phase-encoded in the spin-warp imaging technique (8,9). If N velocity-encoding pulses are applied, of equal duration, but with the magnitude stepped linearly from $(-N/2 + 1)g_v$ to $(N/2)g_v$ then the phase change induced by the nth pulse, in isochromats moving with velocity v is

$$\varphi(n) = \gamma v(N/2 + 1 - n)g_v t_D t_A \qquad [10]$$

where g_v is the incremental strength of the bipolar gradient. If there is a velocity distribution of isochromats $\rho(v)$, then the signal $S(n)$ for the nth velocity-encoding pulse will be

$$S(n) = \int \rho(v)\exp(-i\varphi(n))dv = \int \rho(v)\exp(-iCvn)dv \qquad [11]$$

where $C=\gamma g_v t_D t_A$ and constant phase terms have been neglected. It is thus clear that the discrete N point Fourier transform of S(n) will give a representation of $\rho(v)$ on an N point array.

An example of the practical use of this technique is shown in figure 4, which shows 4 from a set of 8 Fourier flow images obtained in transaxial section through the aorta. A gradient-echo pulse sequence was used, with an incremental velocity sensitivity of 0.24m/s and 8 velocity-encoding steps. The 90° rf pulse occured 70ms after the R-wave, and the total imaging time was 17 minutes. The image matrix was 128 by 128 pixels, the slice thickness 1cm, and the FOV 35cm square. The images clearly show flow in the ascending and descending aorta, pulmonary artery and vena cava.

INFLOW-OUTFLOW EFFECTS

Inflow-outflow effects are the most readily observed flow phenomena, and may lead either to an increase or a decrease in the signal intensity.

Enhancement occurs when TR is less than T_1 of the stationary tissue within the slice, which as a consequence becomes partially saturated. Fluid flowing into the slice has not experienced any previous excitaton, and thus give an enhanced signal relative to the stationary tissue.

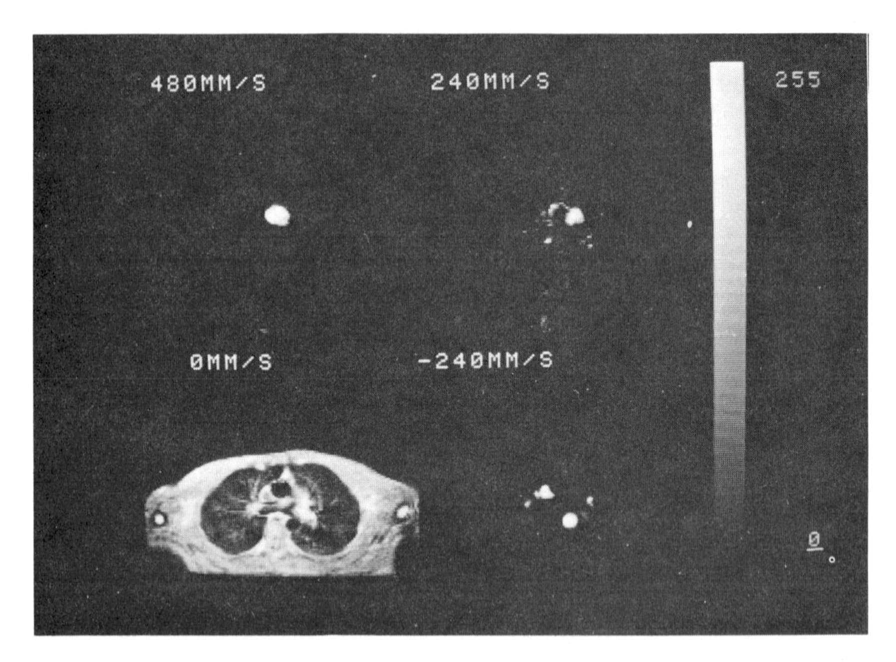

4. Four Fourier flow images showing flow in the great vessels.

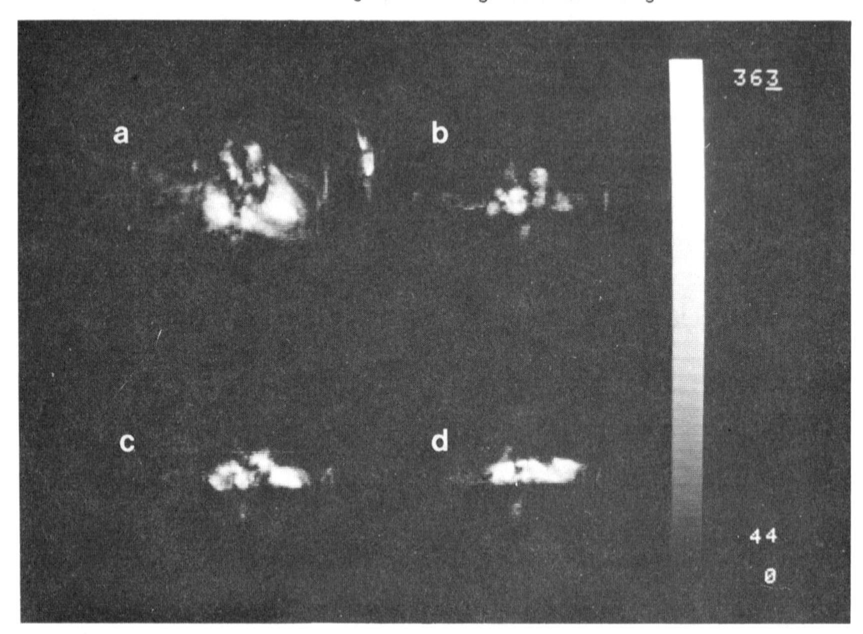

5. Projective spin-echo images through the heart showing time of flight effects.

At high velocities signal loss may occur due to motion in the time TE between the application of the rf pulse and the echo. As the velocity increases this will result in flowing blood having a lower signal level than adjacent tissue.

The quantification of the velocity using this effect is possible if one conducts a number of experiments with increasing TR, (10). The signal from the vessel increases with increasing TR until the portion of the vessel within the slice is totally filled with fresh blood, beyond this point the intensity remains constant. Knowing TR and the slice thickness allows direct calculation of the velocity, if the orientation of the vessel to the slice is known.

This method is now obsolete, due to the long experimental times involved, and its relative inaccuracy. However, flow related enhancement is one of the most powerful qualitative indications of flow.

TIME OF FLIGHT EFFECTS

In the earliest version of this technique two rf coils were placed around the limb of the subject. The downstream coil, which also detects the NMR signal, emits a series of 90° rf pulses. The upstream coil, which is separated from it by a known distance, emits a 180° inversion pulse. At some time later the inverted isochromats reach the downstream coil, modifying the magnitude and phase of the received signal. Thus the velocity may be determined from the time of flight and the separation of the coils.

The imaging techniques that are the extension of this approach employ the selective excitation of two or more spatially separated slices, (11). For example a 90° pulse to an upstream slice, followed some time later by a 180° refocussing pulse to a downstream slice. Only fluid experiencing both pulses will form a spin-echo, and thus only fluid moving within a given velocity range will be imaged.

This technique has the merit that it can image rapidly moving fluids, as there is no restriction that the fluid must remain within a particular 2D slice, but is time consuming and the range of velocities imaged is limited, hence this technique is no longer in common use.

PROJECTIVE TECHNIQUES

Projective techniques dispense with slice selection, and thus no constraint is applied as to the voxel thickness. Hence, the pixels represent the signal from long thin columns of tissue, typically extending through the subject. The flow of blood is then utilised to separate its signal from that of stationary tissue.

The advantage of this approach is that flow information can be obtained from a large volume of tissue thus drastically reducing the examination time in comparison with that necessary for a series of slice selective flow images. A further advantage is that it is possible to follow the convoluted pathways of arteries and veins without difficulty.

Each of the three effects described above has proved to be the basis for at least one method of projective imaging, and these shall now be described in turn.

The first method of projective flow imaging was that of Wedeen et. al (12). The basis of this method is to use to advantage the signal loss due to the dephasing of moving isochromats in a gradient as described above. Two images are collected, one in which the moving isochromats are dephased, one in which they are not. Subtraction of the two images yields a projection angiogram of flow in the direction of the dephasing gradient. This technique may be realised in one of two ways, the first of which is applicable only to arteries. In the case of arterial flow, two images may be collected using identical pulse sequences, one at end systole, when the blood will be flowing rapidly, and hence dephasing, and one at end diastole, when it will be largely stationary, and will not be dephased. The second, and more general method, is to apply two different pulse sequences, one in which the gradients are flow stabilised, the second in which they are deliberately flow sensitized.

The direct application of the phase-sensitive method of flow imaging is impossible to apply when flowing blood constitutes only a small percentage of a voxel, as any motion induced phase change will cause a negligible phase change to the phase of the signal from the entire voxel. However, it is still possible to apply the Fourier method of flow imaging in projective format. This has been termed Projective Fourier Angiography (13), and although more time consuming than Wedeen's method it does have the advantage of being able to determine the direction of flow, and to quantify it.

The two projective methods described above place a heavy load on the ADC (analogue to digital converter), as signal from a very large volume of tissue is acquired, with only a small fraction of this coming from moving blood. One method that seeks to reduce the signal to that from moving isochromats employs the sequence $90^{o}(x)$ - bipolar gradient pulse - $90^{o}(-x)$, or variations on this, with a 180° refocussing pulse between the two 90° pulses (14). The bipolar pulse induces no phase change for stationary isochromats, and so the second 90° pulse tips the magnetization back into the z-direction. Material flowing with such a velocity that the bipolar gradient pulse induces a 90° phase change will not see the second 90° pulse, and will hence give rise to a signal.

Inflow-outflow can also provide projective images (15), although the method is somewhat cumbersome. If a slice is excited with a series of pulses having a very short TR then the signal from stationary tissue will disappear due to its saturation, and only signal from blood flowing into the slice will be visible. The imaging gradients are then arranged so that the slice is viewed edge-on, i.e. the image will consist of a thin band, with the vessels running across the band. A series of such band-images are then obtained from contiguous slices to provide a projective image.

Time of flight effects provide the basis for several projective techniques. The first group rely on pre-inversion of flowing isochromats outside the region of interest. This is brought about either by motion induced adiabatic passage (16), or by use of slice selective inversion (17). The imaging sequence is applied some time after this inversion, during which time the inverted isochromats should have flowed into the region of interest. Again two image data sets are acquired, with and without pre-inversion, and the difference image consists solely of signal from flowing isochromats.

The second method to arise from time of flight effects involves the use of a spin-echo (18). A slice is selected with a 90° pulse, and thereafter a non-selective 180° pulse is applied. Blood that was within the slice for the 90° pulse, but has flowed out of it for the 180° pulse will still form an echo. The imaging gradients are again arranged so that the slice is viewed edge-on, and so the final image will appear as a thin band of signal from stationary material, with excursions from this band due to flowing blood.

An example of this is shown in figure 5. Data were collected on a 128 by 128 matrix, with a pixel size of 3.1mm, and TE=78ms. The subject was a healthy male volunteer. Figure 5a Is included for anatomical orientaion, and shows the major chambers of the heart, the aorta and the pulmonary artery. The slice thickness was 120mm, whereas in figures 5b, 5c and 5d it was 30mm. These images correspond, in sequence, to the top 90mm of figure 5a. The top left excursion of figure 5b shows flow into the aorta, that to the right is into the pulmonary artery, and the lower bolus is flowing down the descending aorta.

CONCLUSION

If NMR flow imaging is to prove clinically successful, then it must offer definite advantages over competing techniques, particularly X-ray contrast angiography and Doppler ultrasound. At present the spatial resolution obtainable is slightly inferior, but acquisition times are comparable with those for contrast angiography. In comparison with Doppler Ultrasound its most obvious advantage is the ability to complement flow information with high quality anatomical images, but this in itself is probably not sufficient to justify the high cost of NMR imaging. As regards contrast angiography, the non-invasive nature of NMR is its greatest asset, and here it may provide useful information on patient who are too ill to undergo catheterisation.

This article has attempted to present the various methods of flow imaging in terms of the physical principles involved, and to show that although a multitude of methods exist they derive from a limited number of effects. It is still too soon to determine which of these methods will be most successful, and indeed what the final relevance of NMR flow imaging will be. The currently favoured techniques are the subtraction angiography methods of Wedeen (12) and Nishimura (17), while the clinical emphasis is on screening patients for cardiovascular disease, and particularly on imaging the carotid bifurcation.

REFERENCES

1. J. M. S. Hutchison, W. A. Edelstein, and G. Johnson, *J. Phys. E* **13**, 947 (1980).

2. G. Johnson, J. M. S. Hutchison, and L. M. Eastwood, *J. Phys. E* **15**, 74 (1982).

3. E. L. Hahn, *Phys. Rev.* **4**, 580 (1950).

4. P. R. Moran, *Magn. Reson. Imaging* **1**, 197 (1983).

5. M. O'Donnell, *Med. Phys.* **12**, 59 (1985).

6. P. van Dijk, *J. Comput. Assist. Tomogr.* **8**, 429 (1984).

7. P. R. Moran and R. A. Moran. In: *Technology of Nuclear Magnetic Resonance*, P. D. Esser and R. E. Johnston, eds. New York: The Society of Nuclear Medicine Inc., 1984.

8. W. A. Edelstein, J. M. S. Hutchison, G. Johnson, and T. Redpath, *Phys. Med. Biol.* **25**, 751 (1980).

9. T. Redpath, D. G. Norris, R. A. Jones, and J. M. S. Hutchison, *Phys. Med. Biol.* **29**, 891 (1984).

10. J. R. Singer and L. E. Crooks, *Science* **221**, 654 (1983).

11. D. A. Feinberg, L. E. Crooks, J. Hoenninger, M. Arakawa, and J. Watts, *Radiology* **153**, 177 (1984).

12. V. J. Wedeen, R. A. Meuli, R. R. Edelman, S. C. Geller, L. R. Frank, T. J. Brady, and B. R. Rosen, *Science* **230**, 946 (1985).

13. D. G. Norris, R. A. Jones, and J. M. S. Hutchison, *Magn. Reson. Med.* **7**, 1 (1988).

14. D. G. Nishimura, A. Macovski, and J. Pauly, *I.E.E.E. T.M.I.*, **MI-5**, 140 (1986).

15. J. M. Pauly, D. G. Nishimura, and A. Macovski, "Sixth Annual Meeting of the Society of Magnetic Resonance in Medicine, New York," p. 28, 1987. [Abstract]

16. W. T. Dixon, N. D. Leila, D. D. Faul, M. Gado, and S. Rossnick, *Magn Reson. Med.* **3**, 363 (1986).

17. D. G. Nishimura, A. Macovski, J. M. Pauly, and S. M. Connolly, *Magn. Reson. Med.* **4**, 193 (1987).

18. D. G. Norris, "Fifth Annual Meeting of the Society of Magnetic Resonance in Medicine, London," p. 19, 1986. [Works in progress].

SESSION IV

APPLICATIONS OF VARIOUS IMAGING METHODS IN ANIMAL SCIENCE

Chairperson: P. Allen

APPLICATIONS OF ULTRASOUND IMAGING IN ANIMAL SCIENCE

Hans Busk
National Institute of Animal Science
P.O. Box 39
DK-8833 Ørum Sdrl.

INTRODUCTION

Ultrasound has been used for many years for measuring on various objects and since the 1950'ies it has also been used for measurement on livestock.

Slaughter quality has always been an important trait in the livestock production and therefore it is included in most breeding goals. There are various systems for measuring slaughter quality on slaughtered animals and the economic settlement for the carcasses is based on quality. It is more difficult to determine slaughter quality on live animals. Great efforts have been made to find suitable methods for measuring of various traits. However, the best suited method so far has been ultrasonic measurement.

The work with ultrasonic measurements has mainly been concentrated on measuring the body composition of the animals. During the last years, however, it has also been attempted to apply the ultrasonic technique in other fields within the livestock production and within research.

Below a short description follows of the ultrasonic technique and those fields of application where it is used to day or might be used in the future. Furthermore, a few results, which are as representative to the various species of animals as possible, are shown.

ULTRASONIC TECHNIQUES

Ultrasound is normally defined as sound waves unperceptible to the human ear e.g. sound waves with a frequency higher than 20,000 Hz (20,000 oscillations/sec.). Due to the high frequency area the oscillations of the sound waves become very small resulting in the fact that the characteristics of ultrasound in several regards assemble light.

An ultrasonic equipment consists in principle of a pulse generator, a sound transmitter, an amplifier together with a cathode ray tube. A certain number of times per second the pulse generator produces a short voltage which in the transducer is transformed into a correspondingly short ultrasonic impulse which penetrates the examined object. If the ultrasonic wave hits an interface between two tissues e.g. from muscle to fat, part of the energy will be reflected to the transducer which also functions as receiver. In the crystal of the transducer, small mechanical oscillations occur which are transformed into electrical pulse. These are amplified and led to the cathode ray valve where they are visualised on the screen. The distance between the measured objects is measured by means of the time interval from when the sound wave is transmitted into the tissue till it is reflected.

The sound wave penetrates a material by putting the single particles into motion. The velocity of the sound in a given material is therefore a function of the elasticity and density of the material.

The velocity of the sound increases typically with greater density of the material. The velocity of the sound is thus approx. 1.450 m/sec. in fat, approx. 1,600 m/sec. in muscle and approx. 3,000 m/sec. in bone. The velocity of sound is furthermore dependent on the temperature.

Determination of the wave length can be made from the following equation:

$$\lambda = \frac{c}{f}$$

where λ = wave length, c = velocity of sound in the medium and f = frequency in Hz. The wave length is of great importance as the least theoretically measureable distance is one wave length. By measurement with a 2 Mz transducer and a velocity of sound of 1,600 m/sec. the wave length will thus be 0.8 mm.

When a sound wave is transmitted into an object the amount of energy reflected from an interface between two materials is dependent on the difference in acoustic impedance in the two materials. The acoustic impedance is defined as the product of the density of the material and the velocity of sound in same material.

The amount of the transmitted energy reflected when the

sound wave hits an interface between 2 materials depends on the difference of the acoustic impedance of the materials and the angle of incidence. If 2 adjoining materials' acoustic impedance is the same, almost the entire ultrasound will pass the interface and no ultrasound is reflected. If, on the contrary, there is a big difference in the acoustic impedance of the 2 materials the ultrasonic wave will be reflected almost completely provided that the ultrasonic wave hits the interface perpendicularly. In order to get as much of the transmitted ultrasound as possible reflected, the ultrasonic wave must hit the interface perpendicularly, see fig. 1. If this is not the case the angle of incidence θ_i will be equal to the angle of reflection θ_r and therefore only a very small part of the transmitted energy is reflected to the transducer. If the ultrasound is not reflected when it hits an interface it is refracted e.g. it continues into the next material. The angle of the transmitted and the refracted ultrasound cohere with the relationship between velocity of sound in the 2 materials according to the following equation:

$$\frac{\sin \theta_i}{\sin \theta_t} = \frac{V_1}{V_2}$$

where θ_i = the angle at the transmitted ultrasound, θ_t is the angle at the refracted ultrasound and V_1 and V_2 are the velocities of ultrasound in the 2 materials.

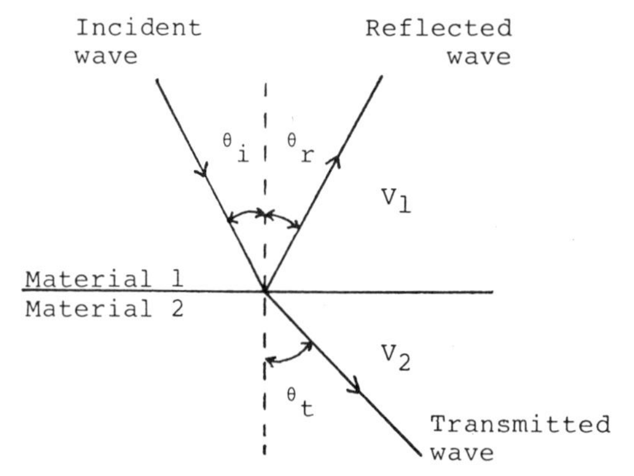

Fig. 1. Reflection and refraction of ultrasound.

ULTRASONIC SYSTEMS

Several different ultrasonic systems are applicable for measurement on livestock. Common for the equipments used for determination of slaughter quality is that they have been developed either to make a depth measurement or to show a picture of a cross section of a specific area of the animal.

The ultrasonic systems can be divided into 2 groups stating how the result of the ultrasonic measurement is presented on the oscilloscop screen, namely the A-mode presentation and the B-mode presentation.

THE A-MODE PRESENTATION

The A-mode presentation visualises the echos as perpendicular fluctuations on the oscilloscop screen (fig. 2).

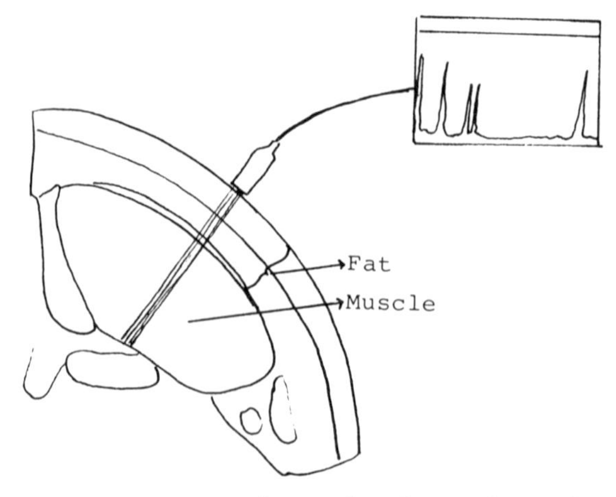

Fig. 2. The A-mode presentation. The fat and muscle depths can be read immediately on the oscilloscop screen.

This A-mode presentation has been used for several years in many countries for measurement of fat depths on livestock. Several experiments have also shown that it is possible to achieve good results with that kind of equipment. It is, however, limited by the fact that it is only possible to measure muscle and fat depths.

THE B-MODE PRESENTATION

By simple B-scanning the transmitter is moved over the examined object without the possibility of variating the angle of incidence of the sound wave. Using a storage screen the echos can be adhered or using a camera turning synchronically with the motion of the transducer (Scanogram) a cross sectional picture of the examined object can be made (fig. 3).

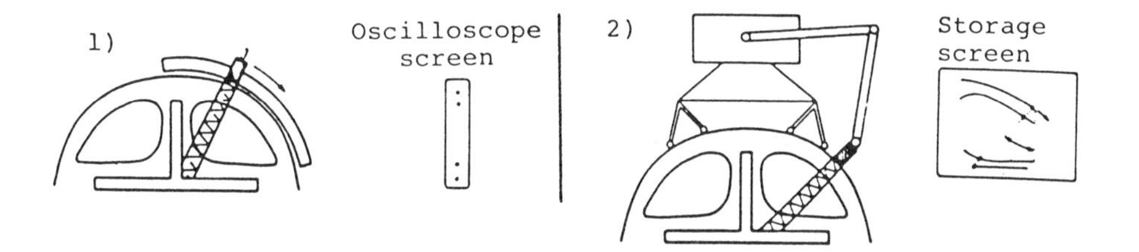

Fig. 3. 1) Simple B-scanning (i.e. Scanogram) and
2) Complexed B-scanning (i.e. SVC-scanner).

By complexed B-scanning the transmitter can be moved both in the horizontal and perpendicular scanning level (SVC Scanner). This makes it possible to make a more perfect scanning picture, as the sound direction more often will be perpendicular on the interfaces. In fig. 3 an example of complexed B-scanning is shown.

REAL-TIME SCANNING

When a transmitter has to be moved over the animal it must necessarily be quiet during scanning. A scan will normally last from 5 to 10 sec. and if the animal moves too much during this period it might be necessary to repeat the scan. When real-time scanning is applied the animal needs, in principle, not be quiet.

A real-time scan means that a scan is made so many times per second (approx. 20) that a firm picture is created on the oscilloscop screen.

The first real-time scanner to be applied on livestock was Siemens' "Vidoson" (Horst, 1971). This scanner uses a single transducer placed inside a house which functions as a reflector. When the transducer rotates a cross sectional scan of the examined object is made.

Newer equipments, e.g. the Danscanner, are real-time scanners which applies another principle than the "Vidoson" by Siemens. All the previously mentioned equipments have used a single transducer. The Danscanner, however, uses a so-called multi-element transducer where the transmitter has 80 small built-in transducers which are electrically connected to each other (fig. 4). When the transducers are activated in rapid succession a cross sectional picture of the object will be shown practically instantaneously on the oscilloscop screen.

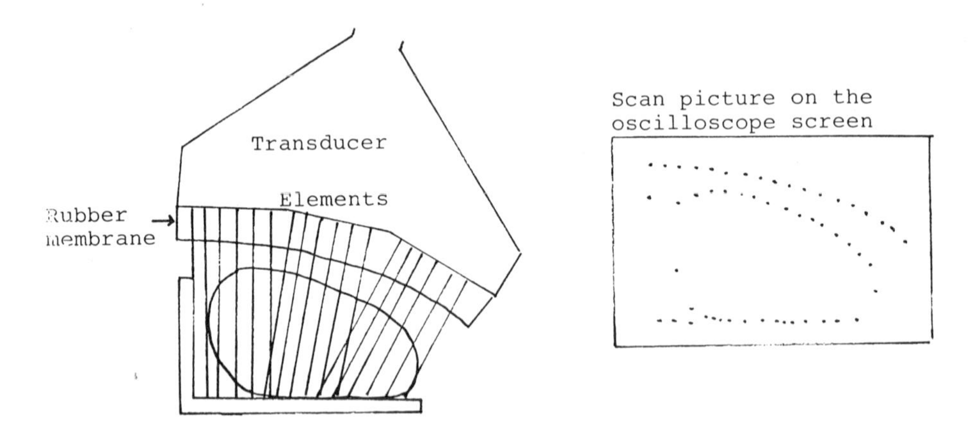

Fig. 4. An example of a multi-element transducer.

Today, multi-element transducers are used for various jobs as they make it possible to make an anatomical scan as well as to follow organic movements.

GREY TONES

As mentioned earlier, the echos can be presented in 2 ways on the oscilloscop screen, either through perpendicular fluctuations (A-mode presentation) or through dots (B-mode presentation). By the B-mode presentation only the strongest echos are shown on the screen as all the weak echos are surpressed in the ultrasonic equipment. In the latest medical equipment, however, grey tones stating the relative level of echos from the various tissues are often used.

The disadvantage of the grey tones is that they often are indistinct i.e. that the interfaces between 2 tissues are not indicated as a thin line at the scanning picture and therefore it will be difficult to make an exact area scan.

MEASUREMENT OF SLAUGHTER QUALITY

Within livestock production ultrasonic measurements have

mainly been used in connection with breeding for determination of slaughter quality on live animals. Slaughter quality is defined differently dependent on species and breeding goal. Within pig production meat percentage in the carcass is the most important trait while dressing percentage, meat percentage and the meat/bone ratio are the most important within cattle production.

With ultrasound some traits are measured, which alone or in combination are correlated to the various slaughter quality characteristics. In many experiments the measured traits on live animals have been compared to control measurements on the carcasses. Such a comparison on e.g. fat depth can be difficult as fat depth might be influenced by slaughter technique among other things. In the following the focus will therefore mainly be on the reliability of the determination of the various expressions of slaughter quality based on ultrasonic measurements on live animals.

PIGS

With pigs there is a good correlation between fat depth and meat percentage in the carcass. In many experiments with slaughtered pigs a correlation between these traits of 0.7-0.8 has been found. Most experiments on live pigs have therefore been concentrated on measuring fat depth.

In several experiments comparisons between the ultrasonic measurements on live pigs and the control measurements on the carcasses have been made (Busk, 1986). The correlation between these measurements is 0.7-0.9 dependent on where on the animal the measurement is made. There are equipments that only measure 2 fat layers and as there are often 3 layers at the last rib this means that the actual fat depth is not always measured. This does, however, not implicate that it is impossible to determine other slaughter quality traits on the basis of the measured fat depth.

As it is normally the meat percentage of the carcass which is to be determined several experiments to prove the precision of this trait through determination by ultrasonic measurement on live pigs have been carried out. In table 1 some results from various experiments are shown.

Table 1. Precision of prediction of carcass traits from ultrasonic measurements in pigs

Authors	Number of animals	Machine	Predict	Measurements	R^2 x100	r.s.d.
Petersson et al., 1982	138	Krautkrämer	% lean	Fd(2)+Md(2)+Td(2)	58.5	
Busk, 1986	133	Renco L.M.	% lean	Fd(2)	63	1.48
	133	Scanmatic	% lean	Fd(4)+Lw	68	1.39
	133	Krautkrämer	% lean	Fd(2)+Lw	73	1.28
	133	Aloka	% lean	Fd(3)+Md(1)	76	1.22
	133	Danscanner	% lean	Fd(3)+Ma(1)+Lw	75	1.22
Kanis et al., 1986	138	Krautkrämer	% lean parts	Fd(4)+Lw	68	1.88
	93	Krautkrämer	% lean parts	Fd(4)+Lw	44	1.83
Sather et al., 1987	55	Krautkrämer	% fat	?	74	1.89
			% lean	?	64	1.69
	55	Scanco	% fat	?	75	1.83
			% lean	?	67	1.63
	55	Renco L.M.	% fat	?	74	1.86
			% lean	?	67	1.62
Stern et al., 1987	200	Krautkrämer	% lean	Fd(3) + Lw	64.7	2.29

Fd = fat depth (number of measurements), Md = muscle depth, Td = total depth, Lw = live weight, Ma = muscle area.

By prediction of meat percentage in the carcass on the basis of one or more ultrasonic measurements on the live pig the coefficient of determination (R^2) is 0.65-0.75 and the residual variation (r.s.d.) is 2.29-1.22. These results correspond to many of the results found on slaughtered pigs when the analogous measurements are applied. In most experiments A-scanning equipments have been applied. In experiments (Busk, 1986) where a comparison of A and B scanners have been made, it has been found that the precision of prediction of meat percentage is only slightly increased by using the more advanced B-scanners compared to the A-scanners.

CATTLE

Due to various breeding goals with cattle there are more definitions of carcass quality. With some beef breeds with great variations in fattness degree the meat percentage is normally the trait to be determined and therefore it will be sufficient to measure fat depth. This can often be done by using A-scanners. With other types of animals, e.g. dual-purpose breeds, dressing percentage and the meat/bone ratio are to be determined, and as the muscle area has a better correlation to these traits, B-scanners should be applied.

In table 2 and 3 some results from various experiments for prediction of the various carcass quality traits are shown (for further information, see Simm, 1983). The results generally show that the muscle area has the best correlation to dressing percentage and to the meat/bone ratio while meat and fat percentages are best expressed through fat depth or fat area.

Table 2. Correlations between ultrasonic measurement and carcass quality in cattle.

Authors	Number of animals	Machine	Measurements	Lean/bone	% lean	% fat
Andersen, 1975	295	SVC	Ma	0.40	0.31	-0.16
			Ma/Fa	0.32	0.53	-0.47
Lykke et al, 1977	31	SVC	Ma(2)	0.56	0.31	-0.17
			Ma/fa(2)	0.61	0.72	-0.65
			Fa(2)	-0.27	-0.66	0.70
Aliston, 1982	50	Danscanner	Fa		-0.66	0.75
			Ma		-0.25	0.37
Andersen et al., 1982	30, 20	Scanogram	Fd (1st lumbar)	-0.38 to 0.09	-0.50 to -0.59	0.54 to 0.67
			Fa "	-0.34 to 0.12	-0.62 to -0.50	0.51 to 0.74
			Ma "	0.33 to 0.62	0.03 to 0.45	-0.31 to -0.03
		Danscanner	Fd "	-0.32 to -0.08	-0.58 to -0.36	0.45 to 0.65
			Fa "	-0.47 to 0.01	-0.69 to -0.44	0.54 to 0.72
			Ma "	0.43 to 0.75	0.23 to 0.54	-0.24 to 0.00
		Philips	Fd "	-0.33 to -0.06	-0.60 to -0.47	0.54 to 0.65
			Fa "	-0.43 to 0.04	-0.72 to -0.43	0.52 to 0.78
			Ma "	0.11 to 0.57	-0.14 to 0.44	-0.23 to 0.20
		Ohio	Fd "	-0.44 to 0.14	-0.65 to -0.44	0.56 to 0.68
			Fa "	-0.44 to 0.05	-0.69 to -0.48	0.57 to 0.73
			Ma "	0.48 to 0.67	0.24 to 0.39	-0.14 to -0.04
Rehben, 1982	60	Danscanner	Ma (2nd lumbar)			-0.31
		R90	Md (3rd lumbar)			-0.21 to -0.04
		Sonic	Md "			-0.27 to -0.18
Andersen et al., 1988	93	Danscanner	Ma (1st lumbar)	0.25		-0.15
		Aloka	Ma "	0.29		-0.09

Miles (1984) has examined whether measurement of velocity of sound through the animal can be related to meat percentage. The advantages by such a method are that the result can be read immediately and that inter- and intramuscular fat as well as subcutanous fat are measurable. By testing of the equipment on live cattle a r.s.d. of 1.34 has been found by determination of meat percentage (Simm, 1983).

The method has also been tested on carcasses (Miles, 1987). It is concluded that the results indicate that it is possible to develop an objective classification based on measurement of sound velocity through soft tissue.

Table 3. Precision of prediction of carcass traits from ultrasonic measurements in cattle.

Authors	Number of animals	Machine	Predict	Measurements	R^2	r.s.d.
Simm, 1983 (Miles)	20	Velocity of ultrasound	% lean	Hind limb + Lw		1.34%
			% fat			1.77 %
	42	"	% fat	"		2.78 %
Aliston, 1982	50	Danscanner	% lean	Fd(2)		2.05%
			% fat	Fd(2)		2.10%
Rehben, 1982	60	Sonic	dressing %	Md	0.52	2.40%
			% fat	Md	0.18	2.90%
		R90	dressing %	Md	0.59	2.20%
		Danscanner	dressing %	Ma	0.67	2.00%
			% fat	Ma	0.40	2.50%

Ma = muscle area, Fa = fat area, Fd = fat depth, Md = muscle depth, Lw = live weight (number of measurements)

Application in cattle breeding

In Denmark measurement with ultrasound for determination of slaughter quality is included as an important trait within breeding. Andersen (1981) states that a genetic improvement of milk and butterfat yield with cattle in the long term will lead to an increase in growth rate and mature size, but also a slight decrease in carcass quality. Import of American Holstein-Frisian for the European Frisian cattle population has also had a negative effect on slaughter quality.

Selection for larger milk yield and import of genes thus contributes to the decrease of slaughter quality of dairy and dual-purpose breeds.

To reduce the negative development in carcass quality, all

bulls in Denmark which are included in the performance test are measured with ultrasound. The measurement is performed 3 times when the buls are 10, 11 and 12 months old, respectively, and the muscle area is corrected to 400 kg after which an ultrasound index is calculated.

PREGNANCY DIAGNOSTICS WITH PIGS

By examination for pregnancy in pigs, A-scanners have been used for several years with variating results. At the examination it is tried whether there are embryo blisters (liquefied blisters) in the uterus. The methods thus only show whether the sow is pregnant or not. The best time for measurement is 30-50 days after fertilization.

Pregnancy can also be determined by B-scanners with a reliability of 100 percent (Inaba, 1983) already on the 22nd day after fertilization. With these equipments it might even be possible to count the number of embryos. Fig. 5 shows a picture of a scan of a pregnant sow on the 21st day. The scan is performed on the side and not rectally as it is done on larger animals. More experiments will be able to decide whether this method is applicable for determination of pregnancy, number of embryos and maybe number of fully developed embryos.

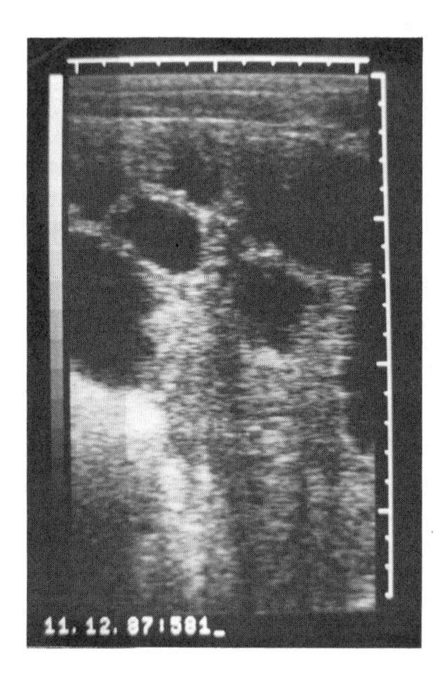

Fig. 5. Picture of a scan of a sow which is 28 days pregnant.

FUTURE USE OF ULTRASOUND

With the new and more advanced ultrasonic equipments it will be possible to use ultrasound for measurement of other traits than slaughter quality. Until now the ultrasonic technique has not been applied very much within research but there is no doubt that certain experiments could be made more efficient if ultrasound could be applied. Some examples are mentioned below.

1. NUTRITION EXPERIMENTS

The reliability of many experiments, among others nutrition experiments, could be increased if it was possible to determine the meat content on live animals before an experiment is started and when it is finished. Morel (1985) has measured fat depth on pigs weighing 26 kg. The pigs were selected from 2 lines, thick and thin fat layer, respectively. The results showed that by measurement with ultrasound a difference of 3.4 mm fat between the 2 lines at 26 kg live weight was established. Jensen (1987) has measured 97 young bulls at an age of 3 and 4 months (early measurement) and at 9, 10 and 11 months (late measurement). The correlation between muscle area and dressing percentage and muscle area and the meat/bone ratio, respectively, was by early measurement 0.45 and 0.20 and by late measurement 0.44 and 0.35. There is thus an indication that it is possible to determine the anatomical composition at a very early stage.

2. SUPPLY OF NUTRIENTS TO THE UDDER

The udder's reception of nutrients can be determined indirectly by measuring the difference in concentration of a nutrient in arterial and venous blood as well as the amount of blood passing through the udder per time unit. By application of ultrasound for determination of amount of blood, measurements can be performed direct on the milk vein without preliminary operation. By means of Doppler shift the velocity of blood in the vein is measured and with another ultrasonic equipment the cross sectional area of the vein is measured. The method has been tried on goats (Jacobsen, 1988). However, more experiments have to be made before it can be concluded whether the cross sectional area of a blood vein can be measured by ultrasound.

3. MINERALIZATION OF BONES

The velocity of ultrasound in a certain material is related to the elasticity and density of same material. It has been tried to benefit from this relationship at measurement of bone quality on horses (Jeffcott, 1985). The results showed that the velocity of sound was less in bones with poor mineralization than in normal bones and that there was a difference of 100 m/sec. in the velocity of sound.

4. VETERINARY RELATIONS

Ultrasound has only been used to a very limited degree for veterinary tasks until now. There is, however, no doubt that it will be used more frequently in the veterinary research in the future.

5. EMBRYO TRANSFER

Diagnostic ultrasound scanners with transducers suitable for intrarectal use have become available for imaging the bovine reproductive tract. Scientists in this field believe that ultrasound measurements has potential in the embryo transfer industry because it can not only be used for detecting the early conceptus and monitoring its viability, but also for evaluating the uterus and ovaries (Seidel, 1988).

Transabdominal ultrasonically guided biopsy has also been tried in an experiment with mares and cows. In preliminary studies, 16 oocytes were collected from 34 follicles in mares and 16 oocytes from 38 follicles in cows.

The ultrasonic measurements will most probably be used to a much larger degree within the embryo technique in the future, as experiments indicate that it could give valuable information about the sexual organs.

CONCLUCION

In the above, examples have been given as to what ultrasound is used for today and where it might be applied in the future.

The advantage of ultrasound is that it is harmless, the equipments are portable, robust and relatively easy to operate. In order to have the full benefit the equipment must, however, be operated by personel with knowledge of the anatomy of the animals. Without such knowledge the measurements would be worthless.

REFERENCES

Alliston, J.C. (1982). The use of a Danscanner ultrasonic machine to predict the body composition of Hereford bulls. Animal Production 35, 361-365.

Andersen, B. Bech (1975). Recent experimental development in ultrasonic measurement of cattle. Livestock Production Science 2, 137-146.

Andersen, B. Bech (1981). Potential use of in vivo techniques in a combined milk/beef breeding programme. Beretning fra Statens Husdyrbrugsforsøg 524, 94-98.

Andersen, B. Bech, H. Busk, J.P. Chadwick, A. Cuthbertson, G.A.J. Fursey, D.W. Jones, P. Lewin, C.A. Miles and M.G. Owen (1982). Ultrasonic techniques for describing carcass characteristics in live cattle. CEC Report, EUR 7640 EN, 78 pp.

Andersen, B. Bech, Per Madsen, Signe Klastrup og Sv.E. Sørensen (1988). Avlsstationerne for kødproduktion 1986/87. 634 report from the National Institute of Animal Scince. Copenhagen, Denmark.

Busk, H. (1986). Testing of 5 ultrasonic equipments for measuring of carcass quality on live pigs. World Review of Animal Production 22 (3). 35-38.

Horst, P. (1971). Erste Untersuchungsergebnisse über den Einsatz des "Vidoson"-Schnittbildgerätes beim Schwein. Züchtungskunde 43 (3), 208-218.

Inaba, Toshio, Yasuhiro Nakazima, Nozonne Matsui and Tatsuo Imori (1983). Early pregnancy diagnosis in sows by ultrasonic linear electronic scanning. theriogemology 20 (1), 97-101.

Jakobsen, Kirsten Christensen (1988). Regulering af næringsstof-forsyning til yveret hos lakterende drøvtyggere. Statens Husdyrbrugsforsøgs årsmøde. 22-25.

Jeffcott, L.B. and R.N. McCartny (1985). Ultrasound as a tool for assessment of bone quality in the horse. Veterianry record 116, 337-342.

Jensen, Merete (1987). Undersøgelse af faktorer af betydning ved ultralydmåling af individprøvetyre. Hovedopgave i kvægets avl. Den kgl. Veterinær- og Landbohøjskole, Husdyrbrugsinstituttet, København, 89 pp.

Kanis, E., H.A.M. van Der Steen, K. De Roo and P.N. De Groot (1986). Prediction of lean parts and carcass price from ultrasonic backfat measurements in live pigs. Livestock Production Science 14. 55-64.

Lykke, Th. and B. Bech Andersen (1977). Use of ultrasonic measurements to predict carcass fatness in cattle. Report from the Royal Veterinary and Agricultural University, Institute of Animal Science and the National Institute of Animal Science, Copenhagen, Denmark.

Miles, C.A., G.A.J. Fursey and R.W.R. York (1984). New equipment for measuring the speed of ultrasound and its application in the estimation of body composition of farm livestock. 93-105. In vivo measurement of body composition in meat animals. (Edited by D. Lister). Elsevier, London.

Miles, C.A., A.V. Fisker, G.A.J. Fursey and S.J. Page (1987). Estimating beef carcass composition using the speed of ultrasound. Meat Science 21. 175-188.

Morel, P. and C. Gerwig (1985). Prediction of meatiness at end of fattening on the strength of ultrasonic measurements at 25 kg live weight. 36th EAAP-Meeting. 10 pp.

Petersson, Harald and Kjell Andersson (1982). Prövning av ekolod och scanningteknik for skattning av köttinnehallet vid individprövning af svin. Konsulentavdelingens rapporter, Uppsala, Sweden.

Rehben, E. (1981). In vivo estimation of body composition in beef. Beretning fra Statens Husdyrbrugsforsøg. 524. Edited by B. Bech Andersen, 48-61.

Sather, A.P., A.K.W. Tong and D.S. Harbison (1987). A study of ultrasonic probing techniques for swine II. Prediction of carcass yield from the live pigs. Livestock Production Science, 14. 55-64.

Seidel, George E., Jr. (1988). Proceedings annual conference of the international embryo transfer society. Theriogenology 29 (1), 3-93.

Simm, Geoffrey (1983). The use of ultrasound to predict the carcass composition of live cattle - a review. Animal Breeding Abstracts, 51 (12) 853-875.

Stern, Susanne, Kjell Anderson, Harald Petersson and Per-Erik Sundgren (1987). Performance testing of boars using ultrasonic measurement of fat and muscle depth. Swedish J. Agric. Res., 17. 47-50.

X-RAY CT FOR BODY COMPOSITION

Odd Vangen

Agricultural University of Norway
Department of Animal Science
Box 25
1432 Aas-NLH, Norway

ABSTRACT

In several research projects, the CT technique has been utilized to measure body composition of live animals and of carcasses. The results have been presented as percent of the total variation described by CT-variables. The accuracy of the developed prediction equations has been tested by crossvalidation techniques and results presented as the ratio between standard error of prediction (SEP) and standard deviation (SD) of the variable.

In PIGS, amount of chemical protein and fat in 208 live slaughter animals was described with R^2-values of 93 and 98 percent, respectively. SEP/SD values down to .30 and .19 were found for the two traits.

In 293 live ram LAMBS of live wt. 42 kg, SEP/SD-values for amount of chemical protein and fat were .26 and .31, respectively.

In CHICKENS, the amount of breast cuts was described with an accuracy (SEP/SD) of .48, while the corresponding figure for amount of abdominal fat was .52.

Over all species, prediction of fat and protein content of live animals was very successfull, with low standard errors of prediction, especially in species with the highest fat content.

As a non-destructive "dissection" method the accuracy of computerized tomography was very high for PIG carcasses (R^2 value of .96 for meat percentage), lower for RAINBOW TROUT (R^2-value of .79 for fat percentage). Based on these promising results, computerized tomography will be utilized in calibration of carcass grading systems and in evaluation of different ultrasonic equipment.

INTRODUCTION

Computerized tomography (CT) is now established as a technique to be utilized in animal research and animal industry. From 1979, when the Nobel Price in medicine was given to the scientists Cormack and Houndsfield for development of the CT-technique in human medicine, a lot of attention was given to this technique and its possibilities in other fields of research. In 1981, Skjervold et al. published the first paper on use of CT in animal science. On the basis of the promising results from scanning of 23 pigs, a computer tomograph was located to the Agricultural University of Norway. From 1982 and onwards several studies on different species have been undertaken, mainly on live animals, but the last years on carcasses and meat products as well.

The present paper will try to summarize these studies and discuss the application of computerized tomography in practical breeding programs, feeding experiments and growth studies.

DATA PROCESSING

Each crossectional picture of a Siemens Somatom 2 scanner consists of 256 x 256 picture elements. In each element, the X-ray absorption (CT-value) is expressed in Houndsfield Units (HU). The CT-values range from -1023 (no absorption) to +1024 (total absorption) and reflects the densities of different tissues and tissue elements. In most body composition studies only CT-values between -200 and +200 were included in the analyses as CT-values of body tissues, except bone, are experienced to be within this range. In some species, like rainbow trout, an even narrower range of Houndsfield Units is utilized in the analyses.

To be able to handle the remaining 400 elements vector describing each scan, frequencies of intervals of CT-values are added together to form the set of independent variables. In most studies intervals of 10 CT-values are used, in other studies, a range of class widths is investigated. Totally the number of independent variables from each scan will normally be 40, but with a variation from 8 to 120 in some studies.

SCANNING PROCEDURES

Normally a standard setting of voltage, scanning width, number of projection and zoom factor has been chosen. The influence of these factors on machine error is studied by Allen & Leymaster (1985). The number of scans needed to describe the body composition in different species and products is variable. In the first studies, 8-15 crossectional pictures were taken. Based on these results, a more limited number of scans are included in the different sets of prediction equations. The costs of additional scans have to be balanced against the possible increase in accuracies.

Live pigs, goats and chicken have to be anesthetized, sheep can be tomographed without. If carcasses or carcass parts are frozen when being scanned, different sets of prediction equations are needed compared to fresh meat, as densities change when freezing and thawing.

STATISTICAL METHODS

Different statistical methods have been used in the development of prediction equations and description of accuracies by use of CT. When up to 40 CT-variables from each of several scans are combined in the same analyses, two main approaches have been applied. The direct contribution of the different CT-variables have been analysed in a stepwise regression procedure. Principal components have also been calculated for each scan by use of eigenvectors. Generally, the results show that the multivariate analysis by use of principal components procedure has lower prediction errors than multiple regression procedures, but the differences between the two methods are very small.

The evaluation of the CT-scanning should not be based entirely on the fitness parameters. Multiple correlation coefficients (or R^2) and standard error of estimate (SEE) are only valid for the data used to construct the equations. Therefore crossvalidation techniques have been used to calculate SEP, standard error of prediction, and the "best" prediction equations are selected on the basis of lowest SEP-values relative to the standard deviation (SD) of the variable. In the present report, comparisons across species are based on this parameter.

RESULTS AND DISCUSSON

Live pigs

Most studies on live animals have been done in pigs. The experiment by Vangen & Allen (1986) included 208 animals of two sexes, representing the normal range of live weights at slaughter. Presented as part of residual variation in body composition from CT-informations of different scans, the comparison to ultrasonics and carcass data is shown in Table 1.

Table 1. Percentage of restvariation explained by CT-informations:

	Protein		Fat		Energy	
	Boars	Gilts	Boars	Gilts	Boars	Gilts
In addition to weight:						
1. scan	47	43	78	85	75	83
1.-10. scan	58	70	95	88	92	83
Best set of CT-values	70	74	96	98	98	97
In addition to weight and slaughter informations:						
1. scan	33	53	73	69	72	70
In addition to ultrasonic backfat thickness:						
1. scan	45	-	87	-	84	-

Vangen & Allen, 1986

From crossvalidation techniques the prediction errors were as shown in Table 2.

Table 2. Prediction of body composition from computerized tomography, live pigs.

	Protein, kg		
	SD	SEP	SEP x 100/SD
Boars	.96	.266	27.7 %
Gilts	1.22	.388	31.8 %
Total	1.11	.506	45.6 %

	Fat, kg		
	SD	SEP	SEP x 100/SD
Boars	1.82	.493	27.1 %
Gilts	3.94	.732	18.6 %
Total	3.17	.585	18.5 %

Vangen & Allen, 1986

As shown from theses two tables, amount of protein and fat is predicted with very low errors. Especially fat content is extremely well explained. Compared to other species, computed tomography is very suitable in studies of body composition in pigs.

A large interest was connected to the use of CT in estimation of meat quality traits, like quality of bacon sides, intramuscular fat content etc. Results are presented in Table 3.

Table 3. Estimation of meat quality from CT-numbers (Allen & Vangen, 1984; Walach-Janiak & Vangen, 1985).

Dependent variable	L.W. + sex	L.W. + S + CT-mean	L.W. + S + CT frequency distrib.
Baconside:			
Percent water	71	87 (55)	-
Percent fat	72	88 (57)	-
Percent protein	67	84 (52)	-
Subcutaneous fat:			
Percent water	59	82 (56)	-
m. longissimus dorsi:			
Percent fat	24	30 (8)	39 (20)
pH	6	12 (6)	20 (15)
Colour	5	14 (9)	25 (21)

( ) Percent of the restvariation in addition to sex and live wt.

As seen from the table, intramuscular fat content is not properly described by CT-variables, while opposite for the quality of bacon side and water content of subcutaneous fat.

The developed prediction equations have been utilized in a study to measure components of growth in pigs during a testing period (22-95 kg). The pigs were scanned 3 times during testing, and protein and fat growth predicted in groups of pigs with 25 and 50 percent Duroc genes. It is concluded from this study (Vangen & Baulein, unpubl.) that CT is an important instrument when measuring changes in body composition during growth. The prediction equations tended to overestimate the total fat content, especially at low live weights. The study showed that protein growth was reduced with increased Duroc-percentage, especially in the last part of the testing period. Fat and energy growth was largest for Duroc crosses, however not increasing from 25 to 50 percent Duroc.

CT is now introduced in the practical pig breeding plan in Norway. All performance tested boars (900 per year) will be CT-scanned, and their own meat percentage will be included as an index-trait.

Live sheep

From a study of 293 ram lambs of the Dala breed, Sehested (1986) developed prediction equations after CT-scanning, slaughtering of animals, dissection of carcasses and chemical analysis of soft tissues. The results from estimating the different body components are presented in Table 4.

Table 4. Predicton of carcass composition in sheep.

Content of	$\bar{X}$	S	R^2x100 and rsd Model 1		Model 2		Model 3		Model 4	
Protein, kg	2.542	.454	82	.190	90	.142	93	.122	94	.112
Fat, kg	2.242	.948	64	.569	89	.323	92	.275	84	.373
Fatfree lean, kg	11.492	2.041	85	.791	92	.571	94	.500	96	.388
Water, kg	8.889	1.569	85	.616	92	.445	94	.398	96	.317
Energy, MJ	147.618	45.677	74	23.043	91	13.628	94	11.759	92	13.179

Sehested, 1986

where model 1 includes body weight.
model 2 includes CT-data from one scan (4.lumbar).
model 3 includes CT-data from four scans.
model 4 includes carcass weight, meat score and weight of kidney fat.

From crossvalidation technique on model 3, residual standard deviations of estimation and prediction were .120 and .137 for kg protein and .275 and .295 for kg fat. In percentage of the standard deviation for the variables SEPx100/SD values of 26 and 31 percent were found for the two variables, respectively. CT-data was regarded as being at least twice as accurate as ultrasound, and use of CT in practical breeding programs may compete with progeny testing in an AI-situation or in a terminal sire system. So far, no efforts has been made to incorporate CT-scanning in the practical sheep breeding plans in Norway. However, this is discussed in a few other countries.

Pregnant and lactating goats

In a Danish PhD-work, Tang Sørensen (1987) did a study in Norway on pregnant and lactating goats. Using the same procedure for developing prediction equations as described for pigs and sheep, changes in water, fat and energy and protein content was measured during lactation and pregnancy. It was found that mainly water and protein was deposited in late pregnancy. All mobilization of energy during lactation derived from fat, and fat from carcass and non-carcass pools were equal labile. During half a year of lactation, the goats mobilized app. 72 percent of their fat, equal to more than 50 percent of their total body energy. Even though the number of animals in this study was limited, the usefulness of CT in these types of studies was demonstrated.

Live broilers

The results of predicting important carcass components in broilers was presented by Bentsen and Sehested (1988). The prediction of amount of abdominal fat and breast cut is presented in Table 5.

Table 5. Accuracy of the best prediction equations for amount of abdominal fat and breast cut in broilers.

	Estimated		Predicted	
	$\sqrt{R^2}$	SEE/SD	r	SEP/SD
Abdominal fat, g				
Regr. model	.90	.444	.84	.570
Princ. comp. model	.89	.469	.86	.522
Sex & L.wt. model	.67	.746	.76	.648
Breast cut, g				
Regr. model	.91	.415	.86	.501
Princ. comp. model	.91	.420	.88	.478
Sex & L.wt. model	.87	.472	.85	.532

Bentsen & Sehested, 1988

It is shown from the table that SEP/SD values were not as low as for pigs and lamb, however, still at very acceptable levels. The most robust CT-observations for prediction purposes were obtained from principal component analysis. Correlations between observed and predicted values for amount of abdominal fat or breast cut were from .7 to .9. In a small scale experiment, it was indicated that in prediction of more direct chemical compositional traits than those analyzed, accuracy of prediction could be further improved.

Slaughtered rainbow trout

Gjerde (1987) published a work on use of CT to predict water, fat and protein percentage of 79 rainbow trouts of 2-4 kg gutted body weight. The results showed that fat percentage could be predicted with SEP/SD-values lower than .5. The author concluded that the method gave high accuracy in prediction of fat percentage. More than one scan position should be chosen and class widths less than 30 HU should be chosen. Compared to ordinary methods of chemical analyses, computerized tomography has high capasity (two persons - 150 fish per working day).

The method is now included in the practical breeding program for rainbow trout in Norway.

Carcass evaluation of pigs

In Table 6 is presented results of accuracies of meat content of half carcasses and bacon sides in pigs (Sehested and Vangen, 1988).

Table 6. Characteristics of prediction equations for half carcass of pigs (N=187).

Dependent variable	SD	RSD	R^2x100
Meat percentage	4.75	1.02	96
Meat content, kg	2.27	.40	98
Meat percent in baconside	6.56	2.22	90

Sehested & Vangen, 1988

It is stated that with the high accuracy of CT-scanning of carcasses, the method is an interesting alternative to dissection of carcasses. The method is non-destructive and it is possible to evaluate 16.7 carcasses per person and day, which is far more than by dissection. The computer tomograph will in future be used in calibration of carcass grading systems and instruments, and is also a tool in evaluation of different ultrasonic equipment.

Meat processing

In a study by Frøystein et. al. (1988), CT has been applied to study salt distribution in hams. The changes in salt uptake and distribution during processing is clearly demonstrated by CT-pictures. The authors are discussing possible extended applications of CT in the studies of meat and other food processing.

REFERENCES

Allen, P. and Vangen, O. 1984. X-ray Tomography of pigs. Some Preliminary Results. In: D. Lister (ed). "In Vivo Measurements of Body Composition in Meat Animals". Elsevier Applied Science Publishers Ltd., London and New York, p. 52-56.

Allen, P. and Leymaster, K.A. 1985. Machine error in X-ray computer tomography and its relevance to prediction of in vivo body composition. Livest. Prod. Sci. 13, 383-398.

Bentsen, H.B. and Sehested, E. 1988. Computerized tomography of chickens. British Poultry Science. (accepted).

Frøystein, T., Sørheim, O., Berg, S.A. and Dalen, K. 1988. Salt distribution in hams studied by computed X-ray tomography (CT). Die Fleischwirtschaft. (in press).

Gjerde, B. 1987. Predicting carcass composition of rainbow trout by computerized tomography. Z. Tierzüchtg. Züchtgsbiol., 104, 121-136.

Sehested, E. 1986. In vivo prediction of lamb carcass composition by computerized tomography. Thesis, Dept. Animal Genetics and Breeding, Agric. Univ. Norway. 81 pp.

Sehested, E. and Vangen, O. 1988. Computer tomography, a non-destructive method of carcass evaluation. Paper, VI World Conference on Animal Production, Helsinki, Finland. 8 pp.

Skjervold, H., Grønseth, K., Vangen, O. and Evensen, A. 1981. In vivo estimation of body composition by computerized tomography. Z. Tierzüchtg. Züchtgsbiol., 98, 77-79.

Sørensen, M. Tang. 1987. Computed tomography for in vivo prediction of body composition in livestock. Ph.D. Thesis, The Royal Veterinary and Agricultural University, Copenhagen, Denmark. 78 pp.

Vangen, O. and Allen, P. 1986. Computed tomography in pig breeding. Submitted to Anim. Prod.

Vangen, O. and Baulein, U. 1988. Computerized tomography to measure components of growth in pigs. (unpublished).

Walach-Janiak, M. & Vangen, O. 1984. An attempt to predict intramuscular fat content, colour and pH of m. longissimus dorsi by means of computerized tomography (unpublished).

X-RAY TOMOGRAPHICAL ANATOMY OF THE SHEEP

A.S. Davies

Department of Physiology and Anatomy,
Palmerston North, New Zealand

ABSTRACT

A continuous series of X-ray tomographs along the head, trunk and hind limbs of a living sheep formed the basis of an atlas and a video production. Disadvantages arising from the unusual position of the animal, and from the inability of X-rays to contrast all tissues and organs, only minimally detract from the ability of computed tomography to depict topographical relationships.

INTRODUCTION

The opportunity was taken, as a cooperative venture between six research organisations, to use an X-ray computer assisted tomographic (CT) machine to determine the body composition of living sheep. "Overfatness" is a common fault in sheep carcasses, and fatness is difficult to estimate in live animals, especially in depots other than the subcutaneous layer. A study of four sheep (Young et al., 1987) showed that CT scanning enables the amount of fat, muscle and bone, as well as their anatomical distribution, to be measured in living sheep. The method has therefore potential in identifying elite breeding stock. Subsequently, a decision was made to use the images from a fifth sheep to produce an atlas of value as a reference work on topographical anatomy (Davies et al., 1987). These images also formed the video production presented at this meeting (Davies, 1987).

MATERIALS AND METHODS

The sheep used was a ewe hogget of the New Zealand Romney breed, about 2 years old, and weighing 56.5 kg. It was anesthetised with sodium pentobarbitone (Nembutal, Abbot Laboratories) and intubated to maintain an open airway for the 5 hours it was unconscious. The sheep was placed in a semi-cylindrical trough, with the forelimbs protracted and the hindlimbs retracted. The CT scanner was a GE 8800 model, manufactured by General Electric, USA. The time taken for each scan was about 2 seconds. A total of 122 slices, each 10 mm thick, was reconstructed. They formed a series from the head at the level of the 4th premolar tooth, to the midshaft to the tibia. Six slices were omitted in the neck region because here the anatomical variation was relatively small. The image was reconstructed in order to contrast fat, muscle and bone to best advantage. The sheep chosen was deliberately fat, with a level of dissectible fat at 34% of carcass weight, so that muscles and other organs would be contrasted by their surrounding fat. Each image was transferred electronically on to X-ray film. These plates were scanned electronically to make plates for printing the atlas, and were also recorded and edited on video tape.

RESULTS AND DISCUSSION

During the exposure time, there must have been respiratory and cardiac movement, but neither of these are particularly apparent. The long period of anesthesia accounts for the distended bladder, and the use of a barbiturate for the large size of the spleen. Although bloating was not a problem, the amount of gas in the stomach and intestine is probably greater than normal. The position of the forelimbs made interpretation of their structures difficult. Body shape, expecially in the abdominal region, was influenced by the shape of the trough in which the sheep lay, and the prone position. The relationships of the abdominal organs cannot therefore truly indicate the situation in the conscious, standing animal. For instance, the left kidney is probably the correct distance from the lumbar vertebrae, but would not lie as close to the abdominal wall when the sheep is standing. The particular levels of contrast used do not distinguish between gas and "raft" (gas, liquid and solid mixture) in the stomach and intestines, and do not show lung structure well; these can be well visualised however, by altering the contrast levels from the electronically stored data. Such manipulation was not, however, sufficient to distinguish structures in certain regions, for example in vessels, lymph nodes and the adrenal glands in the root of the mesentery, and the different parts of the reproductive tract. In the brain, no identification of regions was possible. Bone, cartilage and teeth were generally indistinguishable from one another.

In spite of these shortcomings, many anatomical structures such as individual muscles, lymph nodes, and the parts of the proximal and distal loops of the ascending colon were clearly distinguishable. As a topographical description, the method has a major advantage in that every structure large enough and radiopaque enough is differentiated radiologically from its surroundings. Because there are no gaps between slices, every such structure within a sheep is included in the atlas. The three-dimensional shape of an object can be deducted from the sharpness or otherwise of its outline within the slice. This advantage does not apply to drawings of the surfaces of slices; even the elegant tomographical atlases of human anatomy by Braune (1972) and Eycleschymer and Schoemaker (1911) cannot show three-dimensional detail. The sensation of being able to take a visual journey through a complex object comprising solid, liquid and gas components is especially evident in video production (Davies et al., 1987).

A revolution has occurred in our ability to record, transform, store and reproduce data electronically. This technology has been used here to extend the descriptions of topographical anatomy of small ruminants that began over 100 years ago (Ellenberger and Schaaf, 1884; Ellenberger, 1895), and were continued at intervals by Kolda (1931), Iwanoff (1939), Hopkins et al. (1972), Chomiak et al. (1973), Popesko (1977) and Chatelain (1987). The availability of computer-assisted tomographs, whether derived from X-ray or from magnetic resonance images, is a major advance in the topographical analysis of body structure.

REFERENCES

Braune, W. 1982. Topographisch-anatomischer Atlas. Nach Durchschnitten an gefrorenen Kadavern. Leipzig. Translated into English by Bellamy, E. as: An atlas of topographical anatomy after plane selections of frozen bodies. London, J. & A. Churchill, 1877.

Chatelain, E. 1987. Atlas d'anatomie de la chevre. Capra hircus L. Institut Nationale de la Recherche Agronomique, Paris.

Chomiak, M., Welento, J., Milart, Z. and Szteyn, S. 1973. Atlas anatomii topograficznej zwierzat domowych owca (Ovis aries). Panstwowe Wydawnictwo Rolnicze i Lesne, Warszwawa.

Davies, A.S., Broomfield, N.J. and Povey, T.A. 1987. A journey through a living sheep using X-ray tomography. Video tape, Television Production Unit, Massey University, New Zealand.

Davies, A.S., Garden, K.L., Young, M.J. and Reid, C.S.W. 1987. An atlas of X-ray tomographical anatomy of the sheep. Department of Scientific and Industrial Research Bulletin No. 243. DSIR Science Information Publishing Centre, Wellington.

Ellenberger, 1895. Ein Beitrag zur Lehre von der Lage und Funktion der Schlundrinne der Wiederkäuer. Archiv für wissenschaftliche und praktische Tierheilkunde 21, 62-77.

Ellenberger and Schaaf, 1884. Beitrag zur topographischen Anatomie resp. zum Situs viscerum der Wiederkäuer. Deutsche Zeitschrift für Tiermedizin und vergleichende Pathologie 10, 1-26.

Eycleshymer, A.C. and Schoemaker, D.M. 1911. A cross-section anatomy. New York, D. Appleton & Co., Reprinted in 1970, London, Butterworths.

Hopkins, C.E., Hamm, T.E. and Lepert, G.L. 1972. Atlas of goat anatomy. Part 2. Serial cross sections. Technical report 4426. Maryland, USA, Edgewood Arsenal, Department of the Army.

Iwanoff, S. 1939. Die Topographie der Brustkorbwände und der Brustorgane beim Schafe. Zeitschrift für Anatomie und Entwicklungsgeschichte 109, 544-585.

Kolda, J. 1931. Zur Topographie des Darmes beim Schaf und bei der Ziege. Zeitschrift für Anatomie und Entwicklungsgeschichte 95, 243-269.

Popesko, P. 1977. Atlas of topographical anatomy of the domestic animals. 2nd ed. Philadelphia, W.B. Saunders Co.

Young, M.J., Garden, K.L. and Knopp, T.C. 1987. Computer-aided tomography - comprehensive body compositional data from live animals. Proceedings of the New Zealand Society of Animal Production 47, 69-71.

DISCUSSION

Chairperson: P. Allen/Republic of Ireland

Of the three papers in this session two dealt with the application of ultrasound and X-ray Computed Tomography (CT) mainly for the prediction of body composition in the live animal, whereas the last paper illustrated a very novel application of CT to anatomical mapping of the major tissues in the sheep. This last paper by A. Davies/New Zealand was an excellent demonstration of how new technology can be used as a teaching aid. A continuous series of CT tomographic slices are presented as colour plates in a unique kind of anatomical atlas. By transferring the images to video tape in sequence a "journey" through the body of the sheep can be experienced. This is a most informative and interesting way to learn anatomy and the spatial relationships between the major organs and tissues.

H. Busk/Denmark reminded us that ultrasound has been used as an in vivo method for measuring body composition for a considerable time. The technique is more accurate when applied to pigs than to sheep and cattle. Recent developments include real-time scanning which may lead to improved accuracy. B-mode scanners may be used to measure muscle areas and other dimensions. These are important parameters in themselves in some situations and may improve on the prediction of compositional traits. The relative cheapness, portability and simplicity of many ultrasound scanners suggests that they will continue to have a valuable role to play in live assessment of stock on the farm and at testing stations. The more sophisticated ultrasound equipment is likely to find applications in research. Ultrasound is also used in animal science for pregnancy determination, embryo counting and measurement of the milk vein diameter.

O. Vangen/Norway outlined the results of several years experience of using CT to predict compositional traits in pigs, sheep, poultry and rainbow trout. The degree of accuracy reported is considerably higher than for ultrasound or for that matter any other in vivo techniques, although cost and complexity of the equipment is also on a much larger scale. For this reason the application of the technique is probably restricted to research and large-scale breeding schemes. CT is more accurate in predicting composition in pigs than in sheep, yet it probably has more potential for application in the latter due to the poorer results achieved with ultrasound when applied to sheep. The level of accuracy achievable with pigs is such that it may replace dissection as the calibration method for carcass classification systems and for evaluation of ultrasound equipment. Nuclear Magnetic Resonance (NMR) is unlikely to prove more accurate than CT for gross composition but it has many more potential applications in research due to the greater range of measurements that can be made.

It is obvious from the reviews of the various in vivo techniques that insufficient evidence exists on their relative accuracies. In future trials efforts should be made to include several techniques on a comparative basis

SESSION V

NMR IN ANIMAL SCIENCE

Chairperson: M. Judge

NMR IMAGING STUDIES OF LIVE ANIMALS

MAFoster[1], PA Fowler[1], G Cameron[1], M Fuller[2], CH Knight[3]

[1]Department of Biomedical Physics, University of Aberdeen, U.K., [2]Rowett Research Institute, Aberdeen, U.K., [3]Hannah Research Institute, Ayr, U.K.

1. CONTRAST IN THE NMR IMAGE

In NMR imaging the first objective must be to maximise contrast between the tissues under study. Contrast in the NMR image is mainly derived from proton density and T1 and T2 relaxation times, although other factors such as flow, chemical shift, etc. can help in some cases. Tissues can vary considerably in their basic NMR characteristics and it is, therefore, helpful to have some prior information about their NMR properties which can be used to select an appropriate NMR pulse sequence for the study.

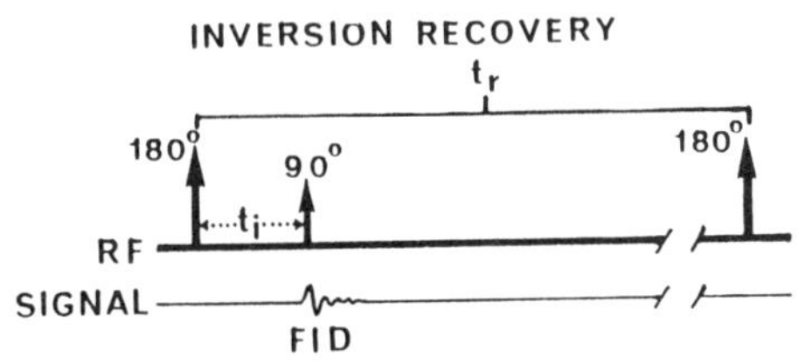

Pulse sequence parameter selection can be illustrated by reference to the inversion recovery sequence (Fig. 1). This involves a preliminary inversion of the spins (180° RF pulse) followed by a 90° slice selection pulse. In the interval (referred to as TI) between these two RF pulses the spins are able to commence relaxation by spin lattice (T1) processes. The signal (obtained by either spin echo or field echo methods), therefore, contains information about the size of the spin population (proton density or spin density) modified by T1 relaxation. The T1 effect will differ between spin populations - a fast relaxing (short T1) population will have advanced further towards the ground state than will a long T1 population and hence contrast is available between these populations in addition to any differences in spin density between them.

There is, however, a complexity in the interpretation of inversion recovery images relating to the relaxation process (Fig. 2).

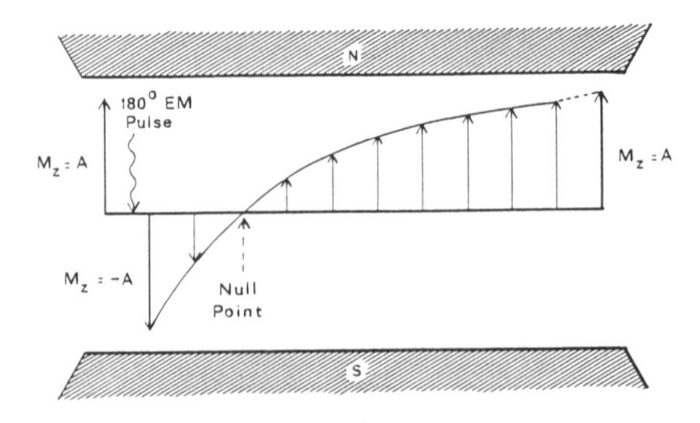

Fig. 2. Signal sign and amplitude examined at progressively greater time intervals after an inverting 180° pulse. After an inverting pulse the spins, in effect, relax back along the direction of the applied magnetic field. They must, therefore, pass through a point (called the null point) at which spin-up is equal to spin-down. In this state it is not possible, after applying a 90° readout pulse, to obtain a signal in the receiver coil. The effect of this null point can easily be illustrated by imagining a test object of a lot of bottles containing solutions of different T1 values. Only one TI is used and if signal intensity in the bottles is plotted as a function of the T1 of the solutions the null point is observed to fall at a slightly higher value of T1 than that equivalent to TI (Fig. 3).

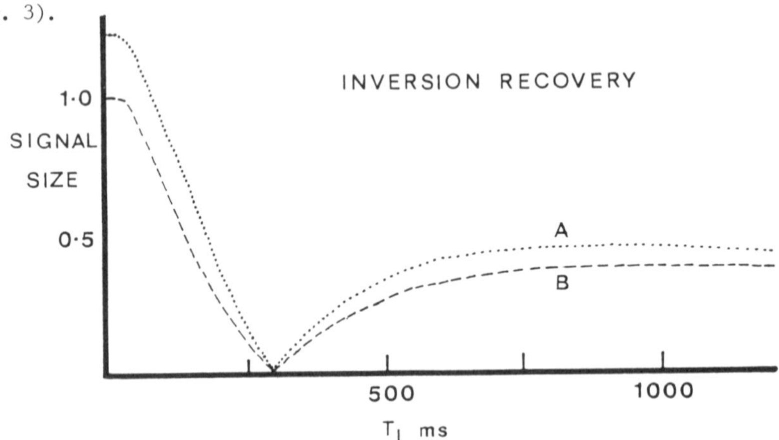

Fig. 3. Signal size vs. T1 value obtained using a standard imaging inversion recovery sequence. TI is 200 ms. Line A was obtained from a series of samples with slightly higher proton density than those used for line B.

Decreasing the proton density of the sample affects the overall magnitude of the curve without affecting the position of the null point. An exactly similar curve can, of course, be obtained using a single sample (i.e. keeping T1 constant) and doing repeated readings with different TIs. Note that, as is the normal practice in imaging (as opposed to NMR spectroscopy), in making this plot only the signal amplitude is considered and not its sign. In effect the part of the curve in Fig. 3 which comes after the null point can be regarded as the negative part of the curve of Fig. 2.

As already mentioned, tissues vary considerably in their NMR characteristics (Table 1) and hence at a given TI value (and given magnetic field strength since tissue T1 is field strength dependent) one can predict the contrast between tissues examined using an inversion recovery sequence.

TABLE 1

NMR proton relaxation times and water contents of tissues from 25-day old rats.

Tissue	T1	T2	%H2O
White Brain	337	89.4	79.0
Grey Brain	392	93.6	81.3
Liver	154	43.5	71.4
Spleen	295	74.3	78.0
Entire Kidney	292	71.5	79.1
Thigh Muscle	256	56.6	78.0
Myocardium	270	59.3	78.4

Values were obtained from excised tissue samples at 2.5 MHz and 30°C. Each value is the mean of eight samples

To illustrate this, in Fig. 4 we see a set of inversion recovery curves obtained using different TI values. Marked onto the lower axis are the T1 values of white and grey brain tissue and of a typical glioma-type brain tumour. The T1 of white brain tissue is shorter than that of grey brain and glioma T1 is much longer than either. Hence on an image collected using an inversion recovery sequence with short TI we will get a lower signal intensity from white matter than from grey.

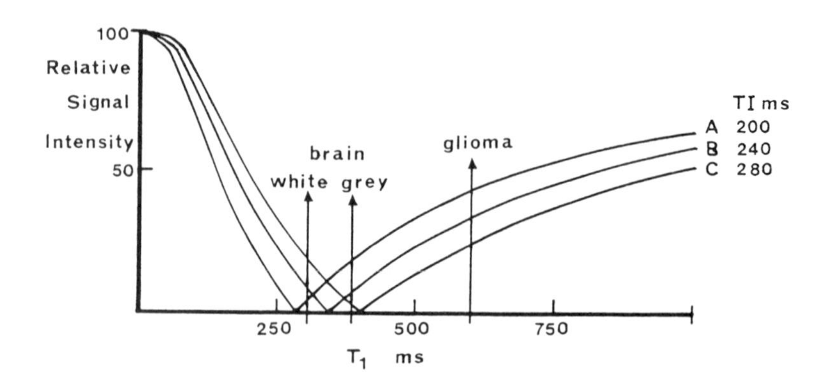

Fig. 4. Signal intensities compared to T1 observed using a short (A), medium (B) and long (C) TI invrsion recovery pulse sequence.

If we choose a TI value which brings the null point between the brain tissue T1 values the signal intensities from the two tissues will be similar (isointense, i.e. no contrast) and both wil look dark on the image. If we use a longer TI interval, bringing the null point to longer T1 values than those of either brain tissue, the white matter will have a higher signal intensity than the grey. By manipulation of this one time interval within the pulse sequence we can, therefore, range from zero to excellent contrast between these two tissues and we can totally invert the image (white darker than grey to white lighter than grey) in the image. Note that the long T1 value of the glioma will always cause it to yield a high signal intensity but the extent of contrast between it and the grey or white matter will depend on the signal intensity of the latter tissues.Greatest contrast between the glioma and normal brain tisses is obtained when a TI is chosen giving a low signal intensity for both normal tissues (line B in Fig. 4). This is just one example of manipulation of pulse sequence to maximise contrast. In, for example, our studies of pig fat we can obtain several times the contrast between fat and muscle using an inversion recovery sequence with appropriate TI than we can with a partial saturation sequence or a T1 map (Fig. 5). Other sequences, e.g. the spin echo sequence which yields T2 contrast as well as T1, can be manipulated in an essentially similar manner.

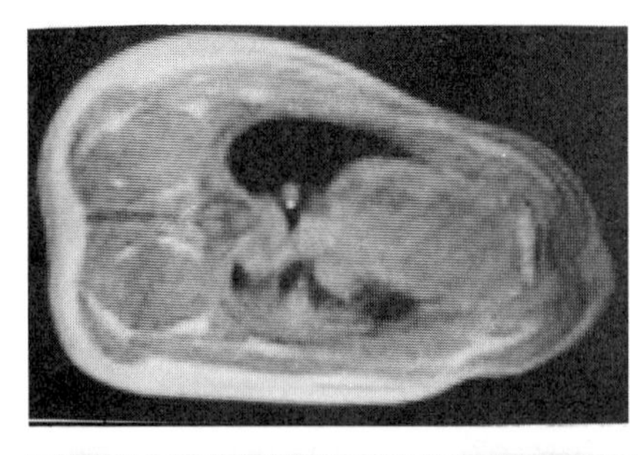

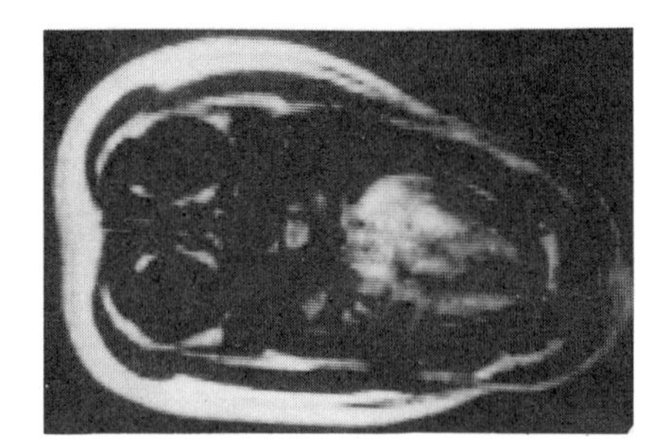

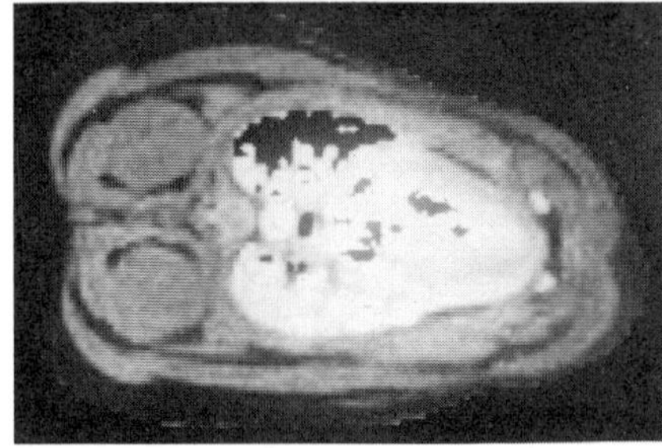

Fig. 5. NMR images of a section through the thorax of a pig showing fat, muscle, spine,heart and lungs. A proton density, B inversion recovery, C T1 map.

2. POST-PROCESSING

Once the basic data of the NMR image is obtained it must be manipulated to enable the best visual impression to be taken from it and, particularly in animal studies, to allow appropriate numerical information to be obtained. This part of the study may be refered to a post-processing. If sufficient fundamental information has been obtained to allow reasonably accurate spin density, T1 and T2 values to be calculated, then a full range of synthetic images can be composed until the ideal one is found. Although this may have uses in certain medical studies where the visual appearence of the image is often more important than its numerical content, in animal studies it is generaly unnecessary and post-processing is limited to defining regions of interest (ROIs) and examining the image pixels (individual image units) within them or, at the most elaborate, to 3-D reconstructions of organs.Most modern imaging sysems allow definition of ROIs, although in some cases these are of severely limited shape. When fully flexible ROI definition is possible they can be used in studies of the texture of tissues (by looking at spread of pixel values within the ROI) for detailed tissue analysis or for generating 3-D models of varying complexity to allow volume estimation. ROI techniques and a simple truncated cone model for volume calculation between slices have been used with considerable accuracy on simple test objects (bottles) and a more complex object (a surgical glove) filled with known volumes of liquid (Table 2). Preliminary results from applications of these techniques to studies of small farm animals and to humans are presented below.

TABLE 2

Errors of valume calculation from NMR images

Test Object	%Error
Simple (bottle)	+4.1
Complex (surgical glove)	+1.1
Excised udder	+3.3

3. APPLICATIONS OF NMR TECHNIQUES

3a. BODY FAT DISTRIBUTION IN THE PIG

We have made some preliminary studies of body fat amount and distribution in the pig. The study is not yet complete but some early results can be mentioned. Pigs (previously fed on standard or enriched diets to produce lean ansd fat animals respectively) were anaesthetised and transaxial NMR images were obtained at thirteen levels down the body. The pigs were subsequently killed and mechanically sectioned at the same places for detailed analysis, this latter process being, as yet, incomplete.

TABLE 3

Differences in body composition between fat and lean pigs

	Pig Type		Significance
	Fat	Lean	(Mann-Whitney)
Overall body area (cm^2) i.e. total area of slices	606.7	602.3	NS
Area of fat (cm^2)	223.7	96.8	<0.001
% fat	35.0	15.6	<0.001
% lean	65.0	84.4	<0.001
Fat T1 (ms)	150.0	157.7	NS
Lean T1 (ms)	248.7	237.3	NS

Data from the NMR results (Table 3) show the ability of the technique to "find" the larger amount of fat on the fat animals (the fat and lean pigs had almost exactly the same total body weight). Examination of the amount of body fat in each of the thirteen sections down the body, however, suggests that there is very little difference in the distribution of fat between fat and lean animals, just in the overall amount of fat (Fig. 6).

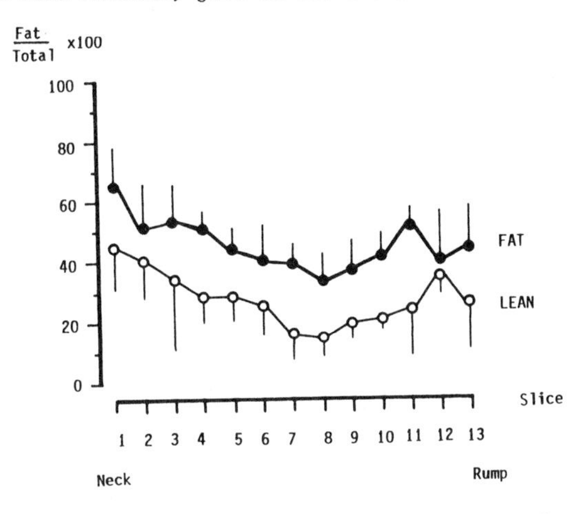

Fig. 6. Proportion of fat to total body area in lean and fat pigs, shown in 13 sections down the body.

The only major differences are in the relatively smaller proportion of fat in the rump sections of the fat pigs compared to the lean pigs. There is, therefore, no evidence for the formation of discrete fat storage sites in the pig.

3b. BODY FAT DISTRIBUTION IN WOMEN

In a study of women pre- and post-reduction dieting we have found that, as in the pigs, the difference in total amount of fat is easily quantifiable from NMR images (Fig. 7). It can be shown that there is a significant change (ANOVA $p<0.001$) in the overall area of comparable sections before and after dieting and (ANOVA $p<0.001$) in the amount of fat in these sections. It is noticeable, however, that the amount of lean remains similar except for the thighs where a loss of lean mass was observed. The greatest loss of fat was from the abdomen.

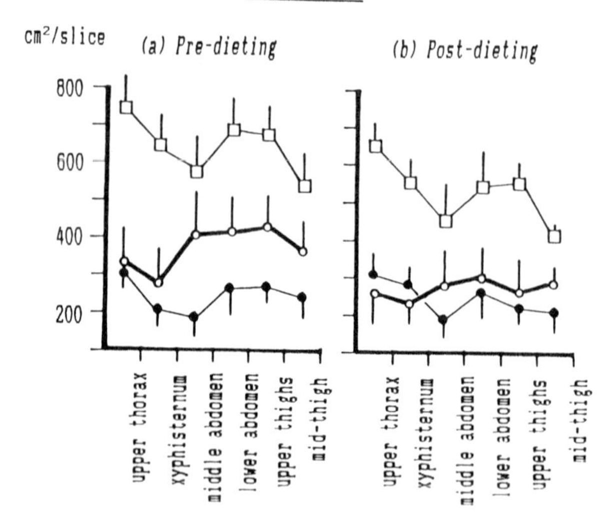

Fig. 7. Distribution of fat and lean in body sections of women pre- and post-dieting.

3c. STUDIES OF THE FOETUS AND NEONATE

NMR imaging is an ideal method for the observation of the foetus in utero (Fig. 8). A pulse sequence can be used with which the amniotic fluid, because of its long T1 value, is highly saturated and hence appears black. This highlights the foetal tissues which have shorter T1 values (though longer than those of the adult).

A major area of interest in our laboratory has been the change in NMR characteristics during tissue maturation. Foetal and neonatal tissues in many species, including the human, have higher water contents and consequently longer T1 and T2 values than their equivalents in the mature animal. Our interest has been in following the pattern of maturation to see if all tissues mature in exactly the same manner and to see if the NMR-observable maturation can be linked with other known biophysical and/or biochemical changes occuring in the tissues. Understanding of the latter associations could provide a better understanding of the general differences in NMR characteristics of tissues.

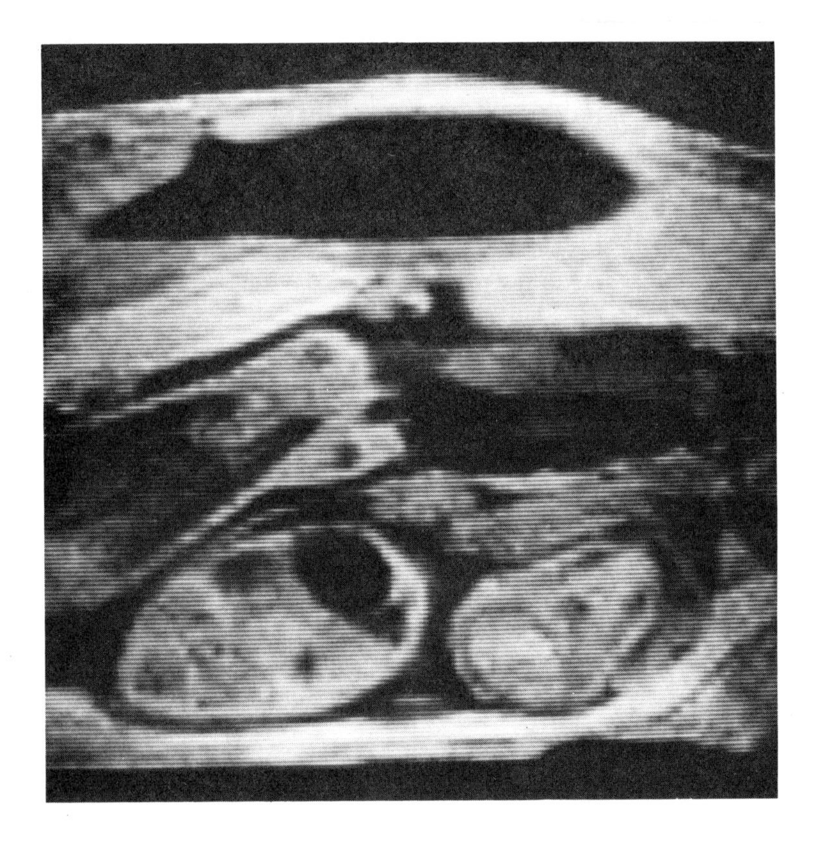

Fig. 8. Transaxial NMR section through the abdomen of a pregnant goat. The upper dark area is the gas-filled rumen. Below this the uterus contains amniotic fluid (also appearing black with this pulse sequence) and two foetal kids. The upper one is seen as an oblique section through the chest showing pectoral girdle and lungs. The fore limbs are extended in front of the body. The lower foetus is bent over and shows as two transaxial sections - one through the head showing nasal sinuses and the brain, the other through the abdomen showing kidneys and the bladder (black).

Most of our work has used the rat as a model and we have found that tissues do vary in their maturation patterns (Fig. 9). For example the liver shows a small decrease in T1 value during the first three to four weeks of life of the rat whilst the muscle, over the same period, shows a much greater change. In some tissues positive links with structural changes can be made, e.g. the white brain, during the first ten days or so, does not show any decrease in T1 or T2 value, then the relaxation times fall rapidly to reach the adult values during the fourth week of life. In the rat the onset of myelination is at about the tenth day post-partum and most myelination occurs within the first four to five weeks. It would seem that a definite association is observed between the NMR properties of white brain tissue and the amount of myelin membrane present.

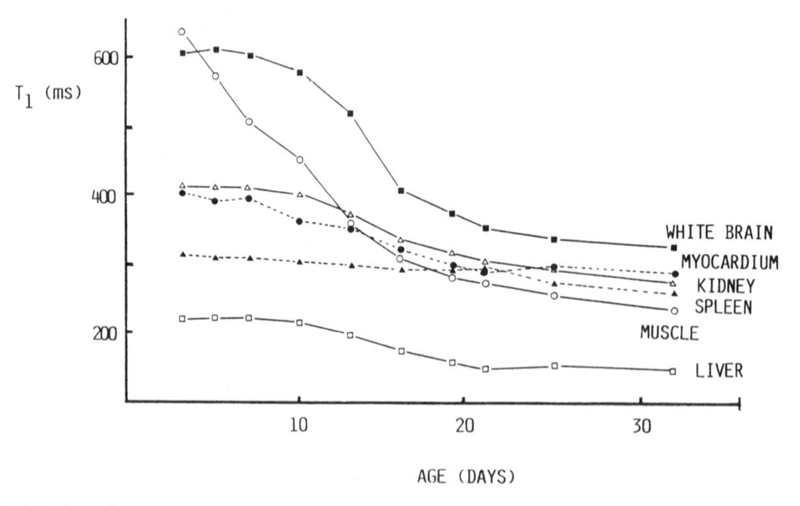

Fig. 9. T1 value variation with age in young rats. Measurements at 2.5 MHz and 30°C.

We have also studied development in young goats using in vitro techniques like those used for the rat study and in vivo methods. The results using both methods were similar and for many tissues were like the general pattern observed in rats. In the goat, however, the changes were less marked in all tissues and especially small for the muscle. A major difference between the rat and the goat is the extent of maturation at birth. The rat is born very immature and feeble whilst the goat is able to rise and walk almost immediately after birth.. It is likely that the differences in muscle maturation pattern observed by NMR reflect differences in the physical or physiological state of the muscle associated with the level of maturity. The guinea pig is also born very mature and able to run about shortly after birth. Here also we see only small post-natal changes in NMR characteristics of the muscle. During late foetal life, however, a sharp decrease in T1 and T2 is observed which is similar to that seen in post-natal life in the rat (Fig. 10).

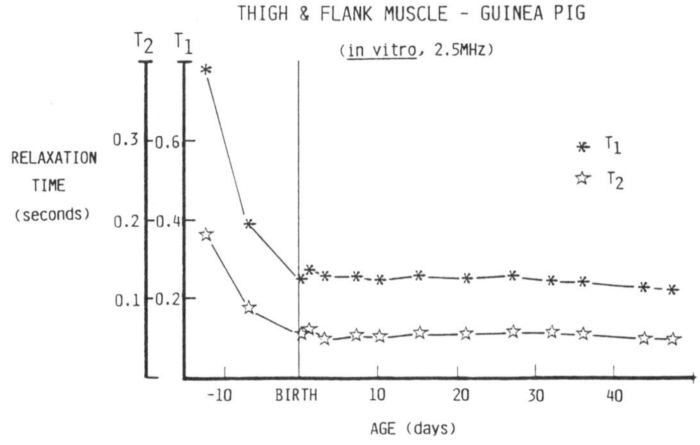

Fig. 10. Variations in T1 and T2 values with age in late foetal and early post-natal life of the guinea pig. Measurements at 2.5 MHz and 30°C.

3d. STUDIES OF THE MAMMARY GLAND

The mammary gland has a very distinctive appearence on an NMR image. In the human mammary tissue is seen diffuse and embedded in the breat fat whilst in an animal such as the goat the udder is discrete and distinct when observed in cross-section (Fig. 11).

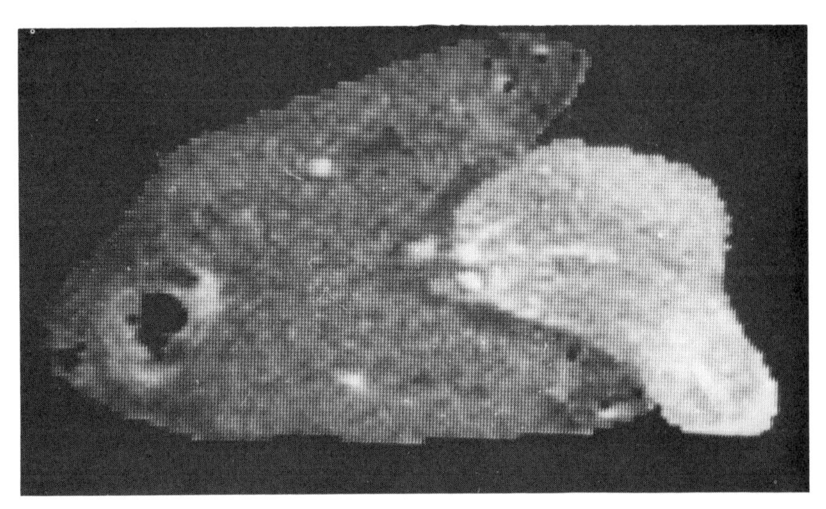

Fig. 11. NMR section (T1 map) through the udder region of a lactation goat.

The udder tissue has a long T1 value, distinguishing it from other body tissues in that region and there is a distinctive fat pad, seen as a darker region between the udder and the general body. Blood vessels, suspensory ligaments and inter-gland septa can be seen in the udder and the milk cistern behind the teat is observed as an area of high signal intensity because of the long T1 value of the milk.

We have been using in vivo NMR imaging techniques to observe changes in the udder during gestation in primi- and multi-parous goats. The goats were sedated or, in late pregnancy, anaesthetised and placed on their sides in the imager. The complete udder region was serially imaged at 2cm intervals and ROI techniques and truncated cone models were used to determine the total areas of the various tissues in each image and the volumes of the whole organ. T1 relaxation time of the organ was also measured. A summary of the results is given in Fig. 12.

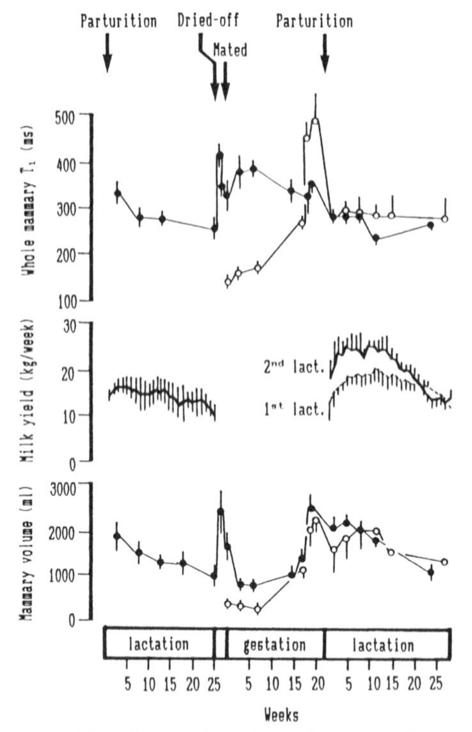

Fig. 12. Summary of milk yield, T1 value and NMR-measured udder volume changes during lacation and gestation in the goat.

During the lactation period in both primi- and multi-parous goats a moderately good correlation (p 0.01) is observed between the udder volume estimated from NMR images and the milk yield (Note - the goats were milked out, with oxytocin aid, before imaging so the udder volume is the volume of the secretory tissue with minimum of milk, not simply a reflection of the milk content of the organ). There is a slightly lower correlation (p 0.05) between T1 value and milk yield.

The large increase in volume of the udder on cessation of milking ("drying off") is observed on the NMR images and is matched by an increase in T1 value of the organ due to the accumulation of the long T1 value milk in the secretory tissues and storage spaces. At this time the udder volume increases by 270% and T1 increases by 160%. The udder volume falls over the next few weeks to reach its lowest at about five weeks into pregnancy. At this time the T1 is still high and this decreases fairly regularly towards parturition. In primi-parous animals, however, the udder T1 value is initially low but shows a marked increase immediately prior to parturition, matching the udder volume increase and undoubtedly associated with fluid accumulation. The volume increase prior to parturition is also observed in the multi-parous animals but is not matched by a large increase in T1 value.

An interesting comparison can be made of the efficiency of milk production in primi- and multi-parous goats (Fig. 13).

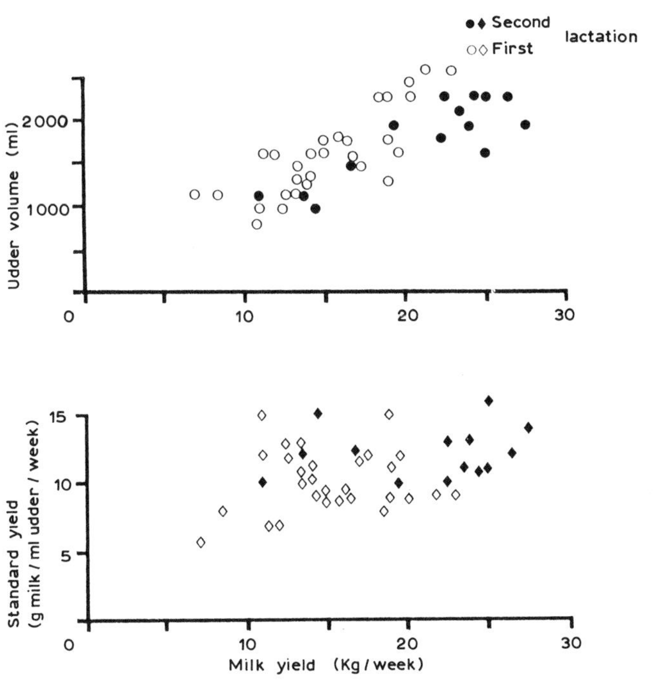

Fig. 13. Efficiency of milk production in primi- and multi-parous goats.

In the upper graph it is seen that goats in their second lactation produce more milk from a lesser udder volume than goats in their first lactation. In the lower graph the same data is shown in terms of the standard yield (i.e. grams of milk per ml of udder per week) compared to milk yield. Here again we observe the greater efficiency of milk production by second lactation goats.

4. FINAL WORD

The four parts from our overall programme described above were chosen to illustrate the range of possibilities for body composition studies available using NMR techniques (NMR imaging and relaxometry only - we do not use NMR spectroscopy at the moment although this also has great potentiallity in body composition studies). NMR imaging is an excellent tool for anatomic delineation especially of fat which has unique NMR properties. The technique can, therefore, be used for measurement of tissue and organ areas and volumes with considerable accuracy. If, however, standard imaging is combined with relaxometry, or relaxation time measurements are made on tissue samples, NMR becomes a powerful tool for examination of biophysical aspects of the tissues. We have only just started to explore the possibilities for applications of NMR methods to body composition studies. Before us is a wide field with a great many opportunities.

SPECTROSCOPY AND MEAT QUALITY

G. Monin and J. P. Renou

Station de Recherches sur la Viande
INRA - Theix 63122 Ceyrat (France)

ABSTRACT

This article deals with the current applications of nuclear magnetic resonance (NMR) spectroscopy in meat science. As compared with the extensive use made of this technique in biochemistry, basic metabolism studies and medicine, NMR has been very little used so far in the study of meat and meat products. However, there is increasing interest in the method, and, since the beginning of the eighties, the number of applications in meat research has noticeably increased. Meat and meat products result from the post mortem evolution of a living tissue, often followed by processing ; so meat science involves many areas of life sciences (biochemistry, metabolism, physiology) and of analytical chemistry. Today, the NMR fields of application relevant to meat science can be listed as follows :
- energy muscle metabolism, intra vitam and post mortem (e. g. mechanisms of halothane sensitivity in pigs, post mortem glycolysis and rigor mortis setting).
- distribution of water in muscle tissue and meat, in relation to waterholding capacity and effects of processing,
- determination of meat composition, mainly contents of fat and protein,
- behaviour of additives such as polyphosphates.

It is concluded that NMR, which may be considered still as being an emerging technique in meat science, has a very promising future.

INTRODUCTION

In 1980, D. G. Gadian gave a review of the potential uses of nuclear magnetic resonance (NMR) in meat science. In his very comprehensive article, he reminded some basic principles of the NMR technique, its advantages and limitations, and suggested a large variety of applications. At that time, NMR had been very little used in meat science, and the author could write : "It is hoped that the examples that are discussed in this chapter will generate enthusiastic interest in applying NMR in the field of meat science". Indeed, not very many studies have been designed using NMR during the eighties, but this is probably due more to the high buying cost of NMR spectrometers or to the difficulty in access to those existing, than to lack of enthusiasm. In biology as well as in analytical chemistry, NMR has proven to be a unique tool of research. Meat and meat products are the results of the changes undergone by living tissues around and after the death of meat animals, and of subsequent processing. As a result, meat science involves many areas of life sciences such as biochemistry, metabolism, physiology and of analytical chemistry. This suggests that NMR is destined to play a great role as a technique of investigation in future meat research.

The aim of this paper is to make a provisional review of the studies which have been designed in the specific field of meat science using NMR spectroscopy. The current NMR applications directly relevant to meat research can be listed as follows :

- studies of muscle energy metabolism, intra vitam and post mortem (glycolysis, rigor mortis setting),
- distribution of water in muscle tissue and meat, in relation to waterholding capacity and changes related to processing,
- assessment of the composition of meat and meat products,
- behaviour of additives in meat and meat products.

Of course, research dealing with protein structure, with relations between structure and function in enzymes -in particular, myofibrillar ATPases and glycogenolytic enzymes (Sykes, 1982)- or with basic muscle metabolism, are of interest for meat science, but are beyond the scope of this paper.

I Muscle metabolism

Nuclear magnetic resonance is a choice method for stuying metabolism both intra vitam and post mortem owing to its non-destructive character. The ability of following metabolic changes during hours on a single preparation avoids artifacts due to muscle sampling and eliminates the imprecision brought about by the variability between preparations. ^{31}P NMR is undoubtedly the most widely used in studies on basic metabolism, physiology and medicine, though other nuclei such as ^{13}C, ^{19}F, ^{23}Na and upmost ^{1}H have met with increasing interest. It is not possible to list here even a small number of the hundreds of works published since the first reports on the possibility of using ^{31}P NMR to study live cells (Moon and Richards, 1973) or muscle tissue (Hoult et al., 1974). Comprehensive syntheses have been written by, inter alia, Gadian (1980, 1984), Barany and Glonek (1982), Bernard et al (1983) concerning ^{31}P NMR, Alger and Shulman (1984) for ^{13}C NMR, Williams et al. (1985) for ^{1}H NMR. The limitation of the NMR technique lies in its relatively low sensitivity, which requires rather large samples (in the magnitude of cm^3), and measurement times sometimes too long for the monitoring of fast processes.

1.1.1. - Studies on live animals and live tissues

To our knowledge, the only studies designed on meat animals or live tissues from meat animals have dealt with the porcine halothane sensitivity. This metabolic defect leads to the well-known pale soft exudative (PSE) meat, of low organoleptical and technological quality (Eikelenboom and Minkema, 1974). Using in vivo ^{31}P NMR, Roberts et al (1983) observed a sudden decrease in muscle phosphocreatine after administration of succinylcholine to halothane - positive piglets. As far as we know, they have not published any further report. We recently found a large difference in sensitivity to muscular fatigue between halothane-negative and halothane-positive piglets using a similar technique (Kozak-Reiss et al., 1988 ; fig 1). Administration of a β-adrenergic agent diminished the difference and enabled the halothane-positive animals to resist muscular fatigue as well as halothane-negative animals. This research is currently in progress.

Kozak-Reiss et al. (1987) used superfused muscle strips to investigate the effects of various drugs on the muscle metabolism of halothane-positive pigs, by means of ^{31}P NMR and mechanical measurements. They observed a fall in phosphocreatine and ATP level under administration of caffeine or of calcium ionophore A 23187 (calcimycine). An original and interesting observation was the development of two intracellular pH compartments in halothane-positive muscle during the development of contracture. This point needs to be confirmed, since such a compartmentation could be artefactual, resulting from defective oxygenation in the middle of the muscle sample due to its relatively large size. This points to one of the limitations of the NMR technique ; because of its rather low sensitivity, large samples are required to reach the short measurement times needed by kinetic studies. One sure way of achieving an adequate tissue oxygenation is the perfusion technique of isolated muscle described by Meyer et al. (1985).

^{13}C NMR can be used to observe glycogen metabolism in muscles of meat animals. The level of muscle glycogen at the time of slaughter determines the meat ultimate pH and is considered as a very important factor in meat quality, in both cattle and pigs. The low time resolution of natural abundance ^{13}C NMR is a major obstacle to the use of this technique on live animals. However the use of isolated muscle perfusion with ^{13}C-enriched solutions could help to overcome this difficulty.

1.1.2. - Post mortem metabolism.

NMR has been applied to the study of post mortem metabolism of muscles from various laboratory animals (for references, see the introduction of the paper by Vogel et al., 1985), but the first research devoted to muscles from meat animals was published by Swedish workers only 3 years ago (Vogel et al., 1985). They analyzed first the post mortem metabolism in bovine muscle, and then the effects of electrical stimulation . In a very detailed investigation, they compared results obtained by ^{31}P NMR and by usual biochemical techniques, and concluded that ^{31}P NMR is a choice method for post mortem metabolism studies. They later reported post mortem changes in pH, lactate, glycogen and phosphocreatine in bovine muscle as studied by ^{13}C NMR and ^{1}H NMR (Lundberg et al., 1986). They discussed the relative advantages of the three techniques, in particular for the intracellular pH measurements. Lundberg et al. (1987) compared ovine and porcine muscles, and

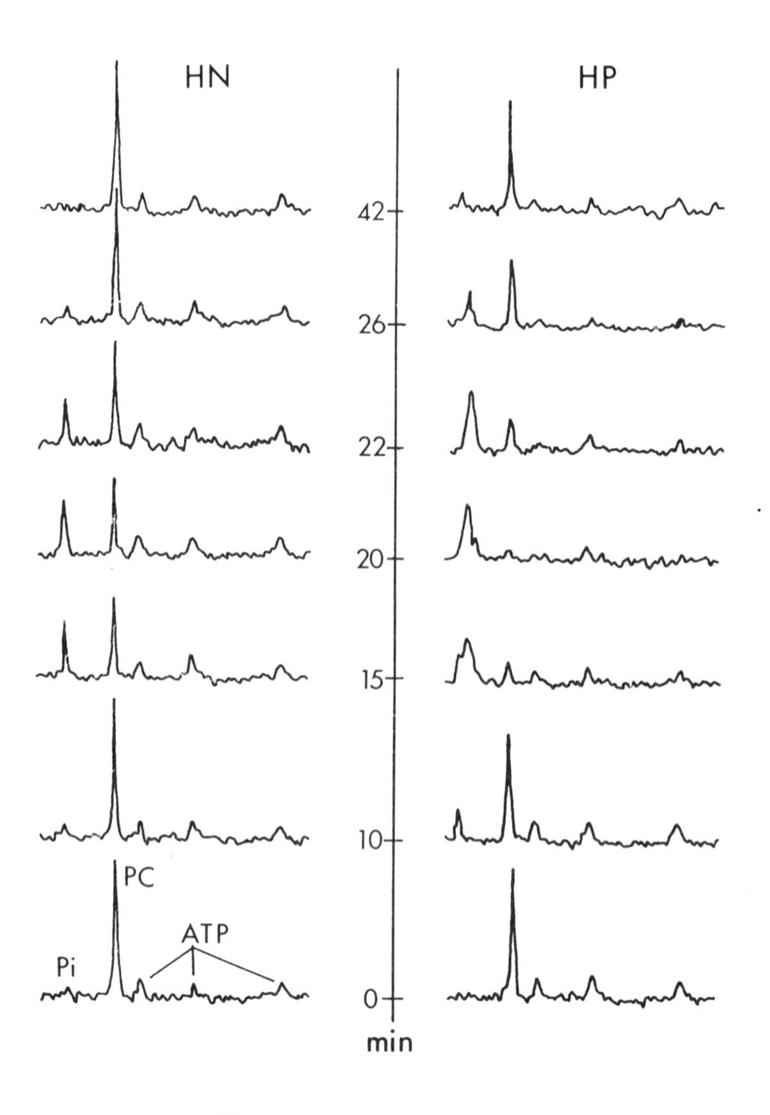

Fig. 1 - Sequence of ^{31}PNMR spectra taken from muscle of live piglets during a muscular fatigue experiment.

Piglets of 30 kg liveweight were anaesthetized using Nembutal Spectra were recorded at 81 MHz for 2 min (32 scans) in a Bruker Biospec spectrometer from the *Semimembranosus* muscle submitted to electrical stimulation using a surface coil. Stimulation sequence: 0.1 Hz from 0 to 10 min; 1 Hz from 10 to 20 min.
HN = halothane-negative ; HP = halothane-positive

studied biochemical changes during thawing of prerigor frozen bovine muscle by ^{31}P NMR. They found that the rate of metabolism varied as follows : thawing bovine muscle > fresh porcine muscle > fresh ovine muscle > fresh bovine muscle.

Post mortem changes in pH and phosphorylated compounds have been studied on rabbit muscles by Renou et al. (1986). They observed differences between metabolic and contractile types. They also found an heterogeneity in the inorganic phosphate signal, which indicates the existence of two pH intracellular compartments, considered to be cytoplasm and sarcoplasmic reticulum. The effect of temperature change between 25 and 15 °C was larger in oxidative - slow twitch muscle than in glycolytic - fast twitch muscle. In chicken muscle, Belton et al. (1987) reported very fast depletions of phosphocreatine and ATP after death.

We recently used ^{31}P NMR to observe post-mortem metabolism in pig muscles that varied largely in rate and extent of post-mortem pH fall (Monin et al., 1988). Some spectra are reported in fig 2. It appears possible to estimate the rate and the extent of glycolysis from the observation of a single spectrum obtained 30 minutes after slaughter : spectrum 1 shows a muscle undergoing slow glycolysis with low ultimate pH, i.e. a normal muscle (high phosphocreatine, low inorganic phosphorus, high pH); spectrum 2 shows a muscle with fast glycolysis and low ultimate pH, i.e. a PSE - prone muscle (low phosphocreatine, high inorganic phosphorus, low pH) ; spectrum 3 shows a muscle with high ultimate pH (low phosphocreatine, high inorganic phosphorus, high pH). Note the very high amount of sugar phosphates in spectrum 2 ; however very high levels were also occasionnally observed in normal muscle.

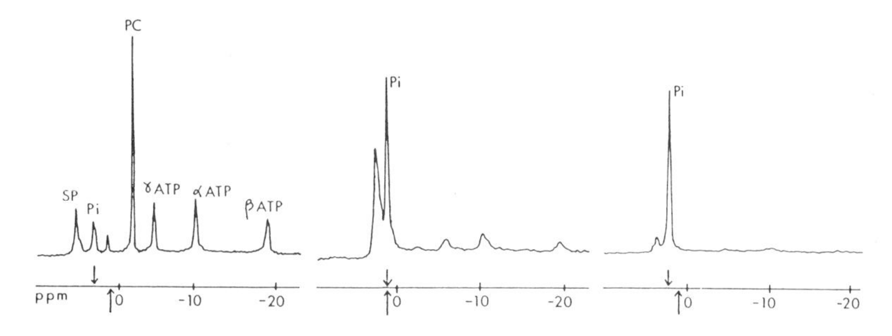

Fig. 2 ^{31}PNMR spectra from pig muscle at 30 min after slaughter.

a. normal muscle : pH 30 min after slaughter (pH_{30}): 6.9 ; ultimate pH: 5.4
b. PSE-prone muscle. pH_{30} : 6.0 ; ultimate pH : 5.5
c. DFD muscle. pH_{30} : 6.7 ; ultimate pH : 6.6

Spectra were recorded on a Bruker AM400 spectrometer at 162 MHz in 10 min by accumulating 224 scans resulting from 45° radio frequency pulses applied at 2.4 s intervals (10 mm diameter probe).

↓ indicates the chemical shift of Pi ; ↑ indicates the chemical shift of Pi corresponding to pH6.

II Behaviour of water in meat.

The water holding capacity and water binding capacity are of major concern for the meat industry, because they affect both sensorial attributes (appearance, tenderness, juiciness) and technological traits (processing yields). Water holding capacity is related to the distribution of water in the meat. Four to 5 % only of the total muscle water is tightly bound at the surface of proteins, the rest being more or less "free" water (Hamm, 1960). Regarding the histological water distribution, around 15 % is extracellular water (variable with metabolic type of the muscle). Most of the intracellular water is retained in the myofibrillar structure and the changes in the myofibrillar filament lattice induced by pH variations, rigor mortis setting or processing (salting, cooking) explain the changes in water holding capacity (Hamm, 1960). Rigor mortis and pH fall induce shrinkage of the myofibrillar network, fibres and fibre bundles. This leads to an increase in extracellular space, particularly large in PSE meat well known for its very low water holding capacity (Penny, 1977). This phenomenon is probably amplified by membrane leakage (Honikel, 1987). The current knowledge about determinism of meat water holding capacity has been fairly well reviewed by Offer (1984).

The behaviour of water in biological samples can be investigated using low resolution ^{1}H NMR. The method is based on the determination of relaxation times of water protons (spin-lattice or spin-spin relaxation). The results give information about the mobility of water relative to the surrounding structures, and the rates of exchange of water between different domains. These techniques have been extensively developed for medical diagnosis, for instance of tumours.

Several authors (Belton et al., 1972 ; Hazlewood et al., 1974 ; Pearson et al., 1974) reported that multiple water fractions are discernible in muscle by measurements of relaxation times of water protons. The changes in the relative size of these fractions during muscle to meat transformation have been studied by Pearson et al. (1974) then Currie et al. (1981). These researchers followed the changes in relaxation times during rigor mortis setting in muscles from meat animals. They discussed their results in terms of water-macromolecular interactions.

Recently Tornberg and her co-workers wrote a series of papers dealing with water distribution in meat and how it was affected by cooking. Tornberg and Nerbrink (1984) demonstrated the suitability of proton relaxation time measurements for the investigation of water behaviour in the myofibrillar structure in various experimental conditions. Tornberg and Larssen (1985) observed that the cooking of beef induced great changes in the domains of water as estimated by relaxation times. They found an increase in "free" water (as determined by NMR) roughly paralleling the increase in cooking loss due to the increase in cooking temperature. They drew some conclusions about the interpretation of NMR data in terms of water distribution, considering that the medium-rate relaxation water was located in the spaces surrounding fibre bundles. In another study dealing with pork cooking, Fjelkner-Modig and Tornberg (1986) found rather different results, i.e. small changes in the water domains during cooking. Moreover the extent of these changes was greatly influenced by breed.

Renou et al. (1983, 1987) used ^{1}H NMR to investigate water-collagen interactions. They found a change in the state of water when the water to collagen ratio increased above 0.35 (w/w). They established that the collagen cross-linking state can be assessed through the cross-relaxation rate between water and collagen protons.

A few studies dealt with the possible use of low resolution NMR as a technique of meat quality assessment. In the previously cited work, Fjelkner-Modig and Tornberg found some relations between relaxation times of water protons and sensory attributes of pork. Correlations although high were generally not significant, due to the small number of samples involved, which makes it difficult to draw conclusions. Using a much larger number of pigs, Renou et al. (1985 b) showed very significant relationships between proton relaxation times or water domains and pork quality traits (Table 1). The level of the correlation coefficients found was promising but still insufficient for precise assessment of meat quality (r 0.7). Borowiak et al. (1986) reported that it was possible to distinguish clearly between normal and PSE pork using spin-lattice relaxation time or a combination of spin-lattice and spin-spin relaxation times of water protons.

There is really a need to establish the meaning of water proton relaxation times in terms of water distribution in muscle tissue and meat. Many researchers assigned fast-rate relaxing water protons to the intracellular space and medium-rate relaxing water protons to the extracellular space. This view is questionable, since Wynne-Jones et al. (1981) have pointed out that there is no need for physical barriers between water domains to give multicomponent relaxation times. Undoubtedly, the ability to use low resolution ^{1}H NMR as a method of meat quality assessment will depend on the progress of knowledge in this field.

	pH 30	pH_u	Reflectance	WHC	Cooking yield
T_1	0.70	0.46	-0.60	0.59	0.35

Table 1. Correlations between spin-lattice relaxation time (T_1) of waterprotons and some quality characteristics in pig *Longissimus dorsi* muscle (from Renou et al., 1985).
pH 30 : pH 30 min *post mortem* ; pH_u : ultimate pH ;WHC : water holding capacity.Correlation coefficients were significant at the $P < 0.01$ level except for cooking yield ($P < 0.05$).

III Meat composition

Fast determination of meat composition is a constant requirement of the meat industry. Manufacturers need to know quickly the contents of protein, fat, collagen and water to guarantee the quality of their products and fullfill legislation requirements. This implies measurements on raw material and on end products. Methods currently in use in the meat industry involve X-ray absorption and near infra-red (NIR) reflectance. For X-ray measurements, the sample must be free of additives ; NIR reflectance is considered to be fast and easy to use once the equipment is well-calibrated, however calibration has proven to be a difficult and tedious task. NMR could be an attractive alternative to these techniques.

Casey and Miles (1974) determined the fat content of freeze-dried meat by low resolution ^{1}H NMR. However, a technique of practical use in industry must be applicable to fresh meat or meat mixtures. Renou et al. (1985 a) proposed a method consisting of determination of proton relaxation times, which was able to measure fairly precisely fat contents between 3 to 40 % on small samples ($1 cm^3$). However, there was some disagreement when the water content varied largely between samples. Renou et al. (1987) therefore developed another method better adapted to industrial purposes, using samples of around 60 g of meat. The technique uses low field - high resolution ^{1}H NMR and allows very fast determinations (a few seconds per sample) in the range 6-60 % fat in ground samples. Unfortunately, at the moment

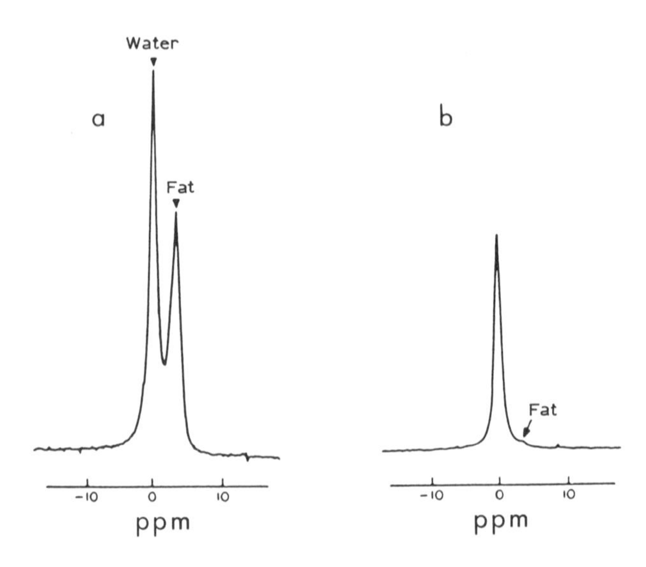

Fig. 3 - Single-scan 'HNMR spectra recorded at 19.6 MHz and 20°C.

a. Ground pork sample, fat content 50.6 % w/w
b. Ground beef sample, fat content 2.5 % w/w

from Renou et al. (1987)

there is no technique to measure fat in fresh muscle (range 1-5 %). Such a technique would be very useful in breeding by making it possible to assess meat composition and potential meat quality in live animals.

Natural abundance ^{13}C NMR provides information on the proportions of saturated - unsaturated fats (Alger and Shulman, 1984 ; Lundberg et al., 1986). This is also of great interest for breeding or feeding experiments, particularly in pigs, since in this species fat composition largely depends on genetic type and on fattening diet.

Other meat components of prime importance such as total proteins and more specifically collagen can also be measured by NMR. Wright et al. (1980) and later Tipping (1982) described a technique to measure meat protein content, which needs little preparation of the sample (grinding and homogenizing in a copper-alkaline reagent). This technique provides a good accuracy and the advantages of safety and greater rapidity as compared with the classical Kjehldahl technique. An apparatus has been specially designed for operating this method (Wright et al., 1980). Josefowicz et al. (1977), then O'Neill et al. (1979) determined the hydroxyproline level, which is generally accepted as the collagen level indicator, in various meat and meat products using ^{13}C NMR. They reported that the observation of hydroxyproline is possible without prior hydrolysis, which is the time-consuming step in the traditional colorimetric technique of hydroxyproline determination. However there is some interference with carbohydrate, whose residual levels in meat may be rather high, which reduces the practical precision of the method. ^{13}C NMR could nevertheless be a method of surveying collagen content at least in certain types of meat (low sugar content).

IV Behaviour of additives

^{31}P NMR has been used to follow the fate of added polyphosphates in meat. Polyphosphates are widely used to improve the water binding capacity of meat and meat products. O'Neill and Richards (1978) then Douglass et al. (1979) observed an hydrolysis of polyphosphates in frozen chicken muscle during long-term storage. It was shown also that fresh breast muscle is able to hydrolyze added polyphosphates for several hours. Recently, Belton et al. (1987 a) reported a detailed study of the fate of polyphosphates in comminuted fresh chicken muscle. They found a great effect of sodium chloride on the rate of hydrolysis, depending on the species of polyphosphate. The same authors investigated the effects of chloride ions and tripolyphosphate on proteins of whole chicken muscle using ^{35}Cl NMR. They concluded that the former cause swelling of myofibrils by increasing inter-protein electrostatic repulsion, without loss of order in the myofibrillar system ; the latter would disrupt the basic protein structure of the myofibrils by breaking cross-links between actin and myosin and so cause a net loss of order (Belton et al., 1977 b).

CONCLUSION

NMR has now proven to be of great interest as an investigation technique in meat research. However, as underlined in the introduction to this paper, it has yet to be used to a greater extent in meat science, at least in high resolution studies. Meat scientists' lack of knowledge of this technique is undoubtedly one reason why it is still little used. The equipment needed for low resolution NMR studies costs roughly the same as that currently used in biochemistry. The unique information it can provide about water domains in muscle tissue or meat has not yet been fully exploited because of insufficient understanding of its import. A large progress margin exists in this field.

NMR spectroscopy may be considered still as being an emerging technique in meat science. The examples of research work already carried out, reported in the present paper, show clearly that large areas of application are open to this technique. NMR spectroscopy is certainly one of the most promising tools for future meat research. Perspectives of use are developed in another paper given in this seminar (Renou and Monin, 1988).

REFERENCES

Alger, J.R. and Shulman, R.G. 1984. Metabolic applications of high-resolution ^{13}C nuclear magnetic resonance spectroscopy. Brit. Med. Bull., 40, 160-164.

Barany, M. and Glonek, T. 1982. Phosphorus-31 nuclear magnetic resonance of contractile systems. In Methods in Enzymology, 85 B, Academic Press, New York, 624-676.

Belton, P.S., Jackson, R.R. and Packer, K.J. 1972. Pulsed NMR studies of water in striated muscle. I. Transverse nuclear spin relaxation times and freezing effects. Biochim. Biophys. Acta, 286, 16-25.

Belton, P.S., Packer, K.J. and Southon, T.E. 1987 a. ^{31}PNMR studies of the hydrolysis of added phosphates in chicken meat. J. Sci. Food Agric., 40, 283-291.

Belton, P.S., Packer, K.J. and Southon, T.E. 1987 b. ^{35} CL Nuclear magnetic resonance studies of the interaction of chloride ions with meat in the presence of tripolyphosphate. J. Sci. Food Agric., 41, 267-275.

Bernard, M., Canioni, P. and Cozzone, P. 1983. Etude du métabolisme cellulaire in vivo par résonance magnétique nucléaire du phosphore-31. Biochimie, 65, 449-470.

Borowiak, J., Adamski, J., Olszewski, K., and Bucko, J. 1986. The identification of normal and watery pork by pulsed nuclear magnetic resonance measurements. 32th Europ. Meating Meat Res. Workers, 9-13, 467-470.

Casey, J.C. and Miles, C.A. 1974. Determination of the fat content of meat using nuclear magnetic resonance. J. Sci. Food Agric., 25, 1155-1161.

Currie, R. W., Jordan, R. and Wolfe, F.H. 1981. Changes in water structure in post mortem muscle, as determined by NMR T_1 values. J. Food Sci., 46, 822-823.

Douglass, M., Mac Donald, M.P., O'Neill, I.K., Osner, R.C. and Richards C.P. 1979. Technical note : a study of the hydrolysis of additives during frozen storage by ^{31}P-F.T. NMR spectroscopy. J. Food Technology, 14, 193-197.

Eikelenboom, G. and Minkema, D. 1974. Prediction of pale, soft, exudative muscle with a non-lethal test for the halothane-induced porcine malignant hyperthermia syndrome. Tijdschr. Diergen., 99, 421-426.

Fjelkner-Modig, S. and Tornberg, E. 1986. Water distribution in porcine m. longissimus dorsi in relation to sensory properties. Meat Sci., 17, 213-222.

Gadian, D.G. 1980. A physico-chemical approach to post-mortem changes in meat-nuclear magnetic resonance. In "Developments in Meat Science" (Ed. R.A. Lawrie), pp. 89-113.

Gadian, D.E. 1984. Phosphorus NMR studies of muscle metabolism. In "Recent advances in the Chemistry of meat" (Ed. A.J. Bailey). The Royal Soc. of Chemistry special publication, 58.

Hamm, R. 1960. Biochemistry of meat hydration. Adv. Food Res., 10, 355-463.

Hazlewood, C.F., Chang, D.C., Nichols, B.L. and Woessner, D.E. 1974. Nuclear magnetic resonance transverse relaxation times of water protons in skeletal muscle. Biophys. J., 14, 583-606.

Honikel, K.O. 1987. Influence of chilling on meat attributes of fast glycolyzing pork muscles. In "Evaluation and control of meat quality in pigs" (Ed. P.V. Tarrant, G. Eikelenboom and G. Monin) (Martinus Nijhoff Publishers, The Hague). pp 273-283.

Hoult, D.I., Busby, S.J.W., Gadian, D.G., Radda, G.K., Richards, R.E. and Seeley, P.J. 1974. Observation of tissue metabolites using ^{31}P nuclear magnetic resonance. Nature, 252, 285-287.

Josefowicz, M.L., O'Neill, I.K. and Prosser, H.J. 1977. Determination of L-hydroxyproline in meat protein by quantitative carbon-13 Fourier Transform nuclear magnetic resonance spectrometry. Analyt. Chem., 49, 1140-1143.

Kozak-Reiss, G., Confort, S., Martin, C., Talmant, A., Monin, G. and Cozzone, P. 1988. A ^{31}P NMR study of muscular fatigue in halothane-negative and halothane-positive pigs (in preparation).

Kozak-Reiss, G., Desmoulin, F., Canioni, P., Cozzone, P., Gascard, J.P., Monin, G., Pusel, J.M., Renou, J.P. and Talmant, A. 1987. Contraction and metabolism traits in skeletal muscle biopsies from halothane-positive pigs as studied by mechanical measurements and ^{31}P NMR. In "Evaluation and Control of meat quality in pigs" (Ed. P.V. Tarrant, G. Eikelenboom and G. Monin) (Martinus Nijhoff Publishers, The Hague). pp. 27-38.

Lundberg, P., Vogel, H.J., Fabiansson, S. and Ruderus, H. 1987. Post mortem metabolism in fresh porcine, ovine and frozen bovine muscle. Meat Sci., 19, 1-14.

Lundberg, P., Vogel, H.J. and Ruderus, H. 1986. Carbon-13 and proton NMR studies of post mortem metabolism in bovine muscles. Meat Sci., 18, 133-160.

Meyer, R.A., Kushmerick, M.J. and Brown, T.R. 1982. Phosphorus nuclear magnetic resonance of fast -and slow- twitch muscle. Am. J. Physiol., 248 (Cell Physiol. 17), C279 - C287.

Monin, G., Talmant, A., Miri, A. and Renou, J.P. 1988. A ^{31}P NMR study of post mortem biochemical changes in normal, PSE and DFD pork (in preparation).

Moon, R.B. and Richards J.H. 1973. Determination of intracellular pH by ^{31}P nuclear magnetic resonance. J. Biol. Chem., 248, 7276-7278.

Offer, G. 1984. Progress in the biochemistry, physiology and structure of meat. 30th Europ. Meet. Meat Res. Workers, 87-94.

O'Neill, I.K., and Richards, C.P. 1978. Specific detection of polyphosphates in frozen chicken by combination of enzyme blocking and ^{31}P-F.T. nmr spectroscopy. Chem. Ind., 1978, 65-67.

O'Neill, I.K., Trimble, M.L. and Casey, J.C. 1979. Carbon-13 pulse Fourier transform nuclear magnetic resonance spectroscopic determination of 4 - hydroxy- L-Proline in meat. Comparison with the colorimetric method. Meat Sci., 3, 223-232.

Pearson, R.T., Duff, I.D., Derbyshire W. and Blanshard J.M.V. 1974. An NMR investigation of rigor in porcine muscle. Biochim. Biophys. Acta, 362, 188.

Penny, I.F. 1977. The effect of temperature on the drip, denaturation and extracellular space of pork longissimus dorsi muscle. J. Sci. Fd Agric., 28, 329-338.

Renou, J.P., Alizon, J., Dohri, M. and Robert, H. 1983. Study of the water-collagen system by NMR cross-relaxation experiments. J. Biophys. Biochem. Met., 7, 91-99.

Renou, J.P., Bonnet, M., Gatellier, P. and Kopp, J. 1987. Calorimetric and NMR relaxation studies on water-collagen interactions. 8th Int. Meeting on NMR spectroscopy. Royal Society of Chemistry, 6-10 July, Canterbury.

Renou, J.P., Briguet, A., Gatellier, P. and Kopp, J. 1987. Technical note : determination of fat and water ratios in meat products by high resolution NMR at 19.6 MHz. Int. J. Food Sci. Technol. 22, 169-172.

Renou, J.P., Canioni, P., Gatellier, P., Valin, C. and Cozzone, P.J. 1986. Phosphorus-31 nuclear magnetic resonance study of post mortem catabolism and intracellular pH in intact excised rabbit muscle. Biochimie, 68, 543-544.

Renou, J.P., Kopp, J. and Valin, C. 1985 a. Use of low resolution NMR for determining fat content in meat products. J. Food Technol., 20, 23-29.

Renou, J.P., and Monin, G. 1988. NMR spectroscopy perspectives. EEC seminar on application of NMR techniques to the body composition of meat animals. Mariensee, FRG, June 13-15.

Renou, J.P., Monin, G. and Sellier, P. 1985 b. Nuclear magnetic resonance measurements on pork of various qualities. Meat Sci., 15, 225-234.

Roberts, J.T., Burt, T., Gouylai, L., Chance, B., Screter, F. and Ryan, J. 1983. Immediate uncoupling of high energy oxidative phosphorylation in muscle of malignant hyperthermic swine determined non-invasively by whole-body 31 P nuclear magnetic resonance. Anesthesiol., 59, A 230.

Sykes, B.D. 1982. Applications of ^{31}P nuclear magnetic resonance to studies of protein structure and function. Can. J. Biochem. Cell. Biol., 61, 155-164.

Tipping, L.R.H. 1982. The analysis of protein in fresh meat using pulsed NMR. Meat Sci. 7, 279-284.

Tornberg, E. and Larssen, G. 1986. Changes in water distribution of beef muscle during cooking - as measured by pulse NMR. 32th Europ. Meeting Meat Res. Workers, 9.4, 437-440.

Tornberg, E. and Nerbrink, O. 1984. Swelling of whole meat and myofibrils - as measured by pulse - NMR. 30th Europ. Meeting Meat Res. Workers, 112-113.

Vogel, H.J., Lundberg, P., Fabiansson, F., Ruderus, H. and Tornberg, E. 1985. Post mortem energy metabolism in bovine muscle studied by non-invasive phosphorus-31 nuclear magnetic resonance. Meat Sci., 13, 1-18.

Williams, S.R., Gadian, D.G., Protor, E., Sprague, D.B. and Talbot, D.F. 1988. Proton NMR studies of muscle metabolites in vivo. J. Magnetic Res., 63, 406-412.

Wright, R.G., Milward, R.C. and Coles, B.A. 1980. Rapid protein analysis by low-resolution pulsed NMR. Food Technology, 34, 47-52.

Wynne-Jones, S., Jones, D.V., Derbyshire, W., Lillford, P.J. and Rodgers, G. 1981. Water proton spin-spin relaxation in muscle tissue. Bull. Magnet. Res., 2, 408.

DISCUSSION

Chairperson: M. Judge/USA

Potential application of NMR in animal science may be limited only by the imagination of investigators. Already demonstrated are applications for determination of body fat and lean in living animals, pregnancy states, functional status of mammary glands, muscle metabolism in halothane sensitive pigs, distribution of water in tissues, composition of meat and behavior of meat additives.

The major NMR parameters most useful in animal science studies are signal intensity, T_1 relaxation times and T_2 relaxation times. Other useful parameters include flow, chemical shift and diffusion events. Demonstrations of methods to vary or optimize contrast have shown the effects of varying the interval between the 180° and 90° pulses (inversion recovery).

Quantitative data processing may be accomplished in different ways. Areas of interest on images may be defined and expressed quantitatively. Volume measurements may be made by three-dimensional reconstruction using areas of consecutive slices and interslice distance.

Spectroscopy methods show great promise for studies of metabolites associated with muscle contraction, postmortem changes in muscle, quality development in meat and fatty acid composition and adipose tissue.

It is possible that NMR will become an important research tool in animal science. A most likely application is its development as a standard reference method to measure body and carcass composition in lieu of dissection methods. On the other hand, some opportunity exists to use less powerful instruments in slaughter facilities to determine carcass composition. Such application would require numerical analysis of images and real-time data capture and analyses.

SESSION VI

CURRENT AND FUTURE PROJECTS

Chairperson O.K. Pedersen

GROWTH PATTERNS AND CARCASS EVALUATION IN PIGS BY MR MEASUREMENTS

E. Groeneveld*, M. Henning**, E. Kallweit**

*University of Illinois Urbana-Champaign, Urbana IL 61801 USA
**Institute for Animal Husbandry and Animal Behaviour (FAL)
D-3057 Neustadt 1/Mariensee FR Germany

ABSTRACT

Quantitative investigations on body composition are of great importance in animal science. Only very few in vivo-methods are applied. Application of MR-imaging is described. Various techniques of image evaluation and generation of quantitative information are explained. An experimental design is introduced which includes in the first stage volumetric measurements of tissues like muscle, bone and fat in live animals and their carcasses, verification of MR measures by total dissection and integral measurements to give a proportion of lean to fat ratio by non-imaging methods. Stage II are in vivo measurements only for tissue alteration (quantitative and qualitativ) during growth under different conditions.

INTRODUCTION

Magnetic Resonance Imaging (MRI) has been developed as a diagnostic tool in the area of humane medicine. It produces images from cross section of the humane body which allows visual inspection by the doctor. The use made from MRI is thus basically an associative process which relies on prior knowledge of the investigating person of what is normal and what is not. A decision derived from MRI in humane medicine usually pertains to one case and could be of the binary type "normal/abnormal". On the contrary, investigations in body composition are basicallly quantitative. The interest lies on the proportion lean in a carcass or a volume or weight of certain portions of the body. Futhermore, a series of measurements on one animal is rarely self contained and self sufficient. If the predictive value of a certain measurement is to be assessed a whole series of images on many different animals has to be considered jointly in a statistical analysis. This has implications in regard to storage and access of image data.
Thus images are only of transient value, we need them as a basis to derive numerical quantitative values on traits we are interested in. This seems indeed to be a mayor difference to applications arising in the diagnostic area.
In the following we shall consider how quantitative statistics can be obtained from MR measurements. This will not only include sta-

tistics derived from images but also discuss the possible use of non imaging techniques.
Lean and fat content of a carcass is of paramount importance in meat producing livestock particularly, in breeding programs. However, this does not provide any information on the location within the carcass or body. Therefore, information on volumetric data on different portions of the body are required in developmental studies. Tissues of interest are muscle, fat deposits and possibly bones. Obviously, spatial information is required in the latter case. Linear measurement can be used as predictors. This again is based on spatial information. We therefore conclude that both statistics are of interest: those which require spatial information and others which do not.

IMAGING

Basically, evaluation of images can be in both dimensions: in a plane and three dimensions. Furthermore, location dependent and independent parameters can be derived.

Two dimensional evaluation

Images are generated on the basis of data matrices whose elements differ in numerical values. High values represent a lighter pixel (a picture element) while darker pixels correspond to lower values. A certain intensity of pixels in an image corresponds again to a certain class of tissue. Muscle is represented by a particular range of pixel values, while fat has a different range. Counting the number of pixels in these two classes should therefore give an indication on the lean or fat content in the slice. However, this might not work for various reasons. Contrary to X-ray CT there are no standardized values in MRI. The numerical values of pixels are subject to changes in the measurement parameters and conditions which may change from one measurement to the next. Seemingly, this problem is difficult to overcome. It might be possible to choose a known tissue like eye muscle as a standard. However, if the variance of pixel values is different from one measurement to the next, this will not help much either. Once standardization of pixel values allows grouping of pixel values according to the type tissue they represent, regression analysis can be employed to derive prediction equations for body composition along the lines the Oslo group around Skjervold developed.

Location dependent

If lean in different cuts is of different importance, analysis of images has to consider location information. Growth studies investigating differential development of tissues require this approach.

A very straight foreward method is **manual delineation** of images. Image processing software is usually part of MR imagers and may produce values for areas. This is, however, very cumbersome for large numbers of images.

In this case image processing techniques can be of help. The first step to automatic identification of groups of tissues such as eye muscle is **edge detection.** As we have seen above absolute values of pixels may be different even for the same tissue. However, changes of pixel values as we go from one type of tissue to the adjoining are fairly stable. It is these changes of values on which edge detectors are based. The process of edge detection can be visualized as scanning rows and columns of an image, noting regions of maximum change. This maximum is most likely a border between different tissues. All surrounding values, except for the maximum, are deleted resulting in an image which should ideally represent borders amongst tissues only. However, the matter is complicated by artefacts, and also by random variation of pixel values, also called noise. Filters can help in this regard which basically smooth the image. A careful choice of thresholds may reduce the risk of detecting non existing borders.

What is very easy for the humane eye, might be difficult for a computer. While arithmetic operations can be handled with ease on computers, any type of association is amazingly difficult to achieve. While every child can follow a broken line on a picture and recognize the pattern of e.g. a kidney this is very complicated on a computer. One approach to tackle the problem is to allow limited humane intervention, for instance, by positioning a curser inside the eye muscle area. The software could then perform what is called **region growing**. Starting from the pixel the cursor was positioned on, the program would scan the surrounding area, thus "growing a region" around the starting point. The process would be terminated when the previously generated edge is detected. Another approach would be to either follow previously defined edges or to follow drastic changes in pixel values in any direction within the two dimensional space of an image. Apart from being computionally expensive the problem of broken lines has to be considered, since this might lead to a loss of track.

Three dimensional evaluation

Evaluation of images is only the first step to determining parameters of interest like estimates of volume, amount or weight of certain classes of tissue. Basically two approaches are available to establish estimates of three dimensional samples.The first employes regression techniques to predict the three dimensional sample from values derived from two dimensional images. This is basically done when eye muscle area is used as a predictor of lean in the carcass with the difference, that MR measurements can be taken on live animals. Different transversal slices will have different predictive power. Finding the optimum number and position of slices is a matter of investigation. If, however, the complete body is scanned, it should be possible to calculate, not estimate, the proportion of different tissues. Based on the classification of tissues in single images the tissue classes can be accumulated as more slices are added.

While this process could be called **pseudo three dimensional**, MRI can generate a truly **three dimensional** data set consisting of a 128*128*128* or 256**3 data space. Such a data set would not suffer from border line problems as occur in a series of slices. However, the volume of data is formidable: One three dimensional data set of dimension 256 comprises 48 Mega bytes!

INTEGRAL MEASUREMENTS (ONE SHOT)

A MR experiment conducted at the proton frequency produces a signal whose amplitude at time 0 is proportional to the amount of protons in the sample. Muscle and fat both contain protons, muscle in form of water, fat in form of CH groups. The different chemical binding and surrounding result in different shapes of the signal decay curves. At a higher magnetic field fat and water produce two distinct peaks which are 3.5 ppm apart on a 1.5 tesla spectrometer. At lower fields, however, they are not separable. This results in one signal decay curve which is based on the sum of the two components fat and water. The spin echo relaxation for a "pure" sample like water follows an exponential function, that of a sample containing two components a double exponential function, which is the addition of the two separate "pure" curves. If the two components are different in respect to the time constant (i.e.

the time it takes until the signal fades away) it is possible to extract the two underlying component functions. If one component has a short half life in comparison to the other, the last part of the curve will be based only on the long component. Fitting a regression line to this portion of the curve and extrapolating to time 1 gives an estimate of the complete curve of the "long" component. **Graph 1** shows signal decay curves from the ham , the neck and the brain area from a live pig. The different proportion of lean and fat in the three areas result in different shapes of the curves. Extrapolation of the separated curves at the true time 0 gives an estimate of the proton density of fat and water in the sample. Considering that muscle contains 80% water while fat contains around 17%, the ratio of the intensity of water $I_{(H2O)}$ to that of fat $I_{(fat)}$, should be an estimator of the proportion lean in the sample. If the MR experiment is run at higher magnetic fields fat and water curves are produced separately from the beginning. The unit we have been talking about so far has been a "sample". Ideally a complete pig should be contained in the sample. Since this is not possible (at least not at higher weights) we have to consider the relationship between an sample and a complete animal. On a magnetic resonance imager a sample can be defined both conceptually and operationally as a slice from which an image is produced. While this requires three gradients only one slice gradient is required in a spectroscopic experiment as described above. Results obtained from an imager pertain the slice selected and give an estimator of the ratio of fat and water in the slice. However, we are interested in parameters like kg of lean and kg of fat in the complete pig.

First it has to be noted that the MR experiment does not provide a measure in absolute terms, since the signal intensity does not only depend on the thickness of the slice but also on a variety of measurement parameters. Therefore, they are not reproducible and cannot be used on an absolute basis. One obvious solution would be to use the MR measures as auxiliary traits and employ regression techniques to predict whatever we are interested in. In doing so we concede that we know nothing about the causal relationships between the auxiliary trait and the dependent variable. We would take MR measurements at various positions on the body, establish the true body composition by total dissection and use stepwise regression techniques to assess the predictive value of various sli-

Integrated Signal Decay Curves, 16 Echos

Transversal Slices

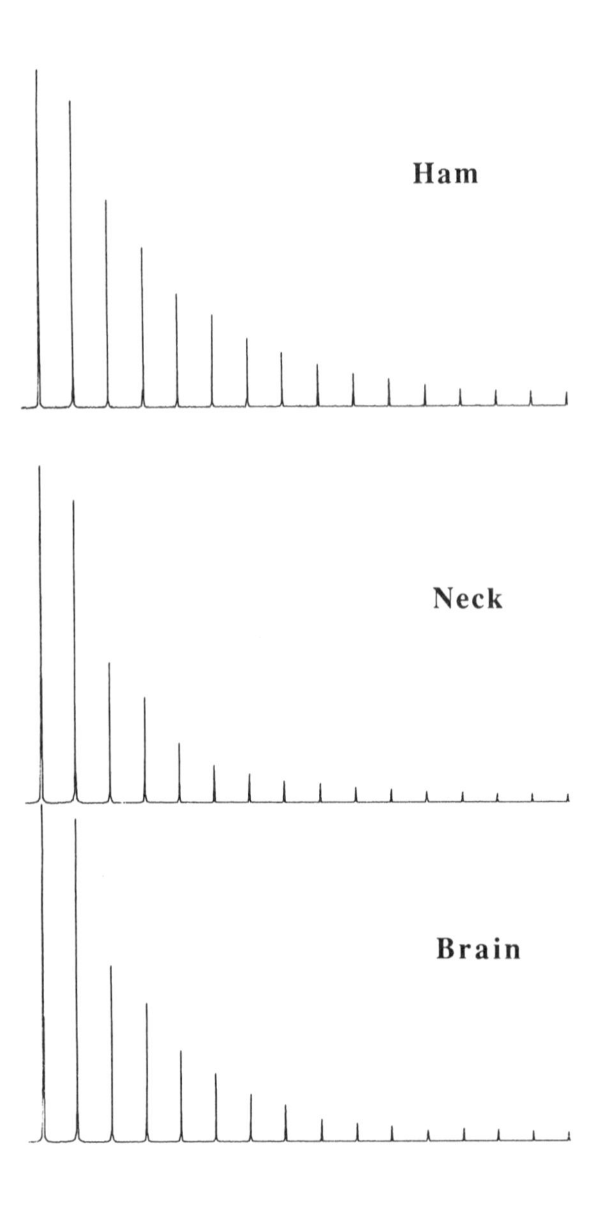

Graph 1

ces, and finally establish a prediction equation. This is a perfectly legitimate procedure, and will probably produce good results. However, it seems that MR spectroscopy measures much closer related to what we are after than all the other techniques so far: Within the slice we should get a rather precise measurement of the proportion of fat relative to water. This measurement is not based on only a one dimensional measure but rather on all the fat and water contained in the slice. In regard to the sample, it is, therefore, more than an estimation procedure along the line pars pro toto. Since water is a constituent of muscle, blood and intestines and basically in every tissue including bones, it seems impossible to determine the volume of muscle and fat on the basis of the ratio of their intestines: we have two unknowns but only one equation. On the other hand, if it was possible to use the signal intensities for water and fat directly, for instance by keeping measurement parameters constant and including the volume of the sample via counting the number of pixels and multiplying this by the thickness of the slice, two equations could be generated. Unfortunately, this implies image reconstruction which we would like to avoid. It, therefore, seems that regression techniques have to be used to predict the amount of lean and fat in the carcass. Since the measurement time is very short, for 16 echoes with an interecho time of 32 msec only half a second, basically two routes are available to improve overall prediction accuracy. Firstly, more than one measurement at the same location can be carried out, thereby improving signal to noise ratio. From the **graph** 2 it seems that already one measurement is fairly accurate, there does not seem to be much random variation (the second data point is an outlier) which is a known deficiency of the measurement technique. Scanning more slices of the body seems to be a more promising way to increase accuracy of prediction.

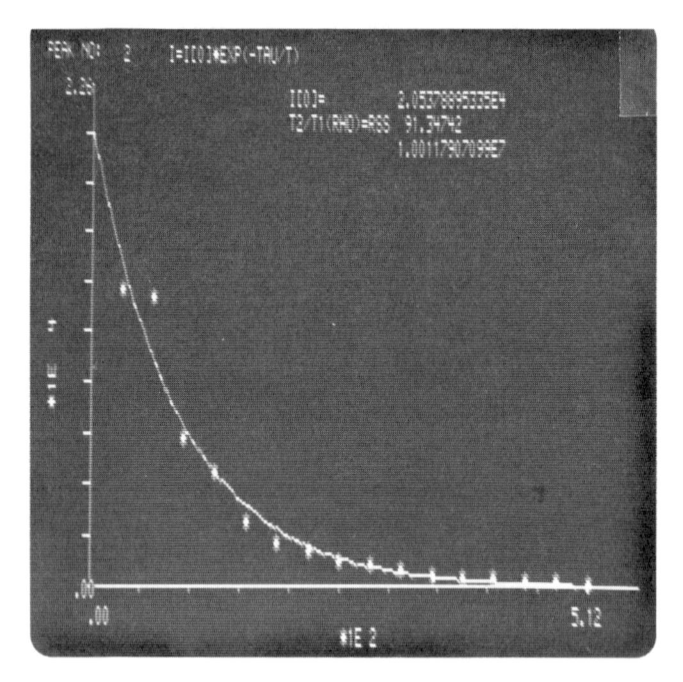

Graph 2

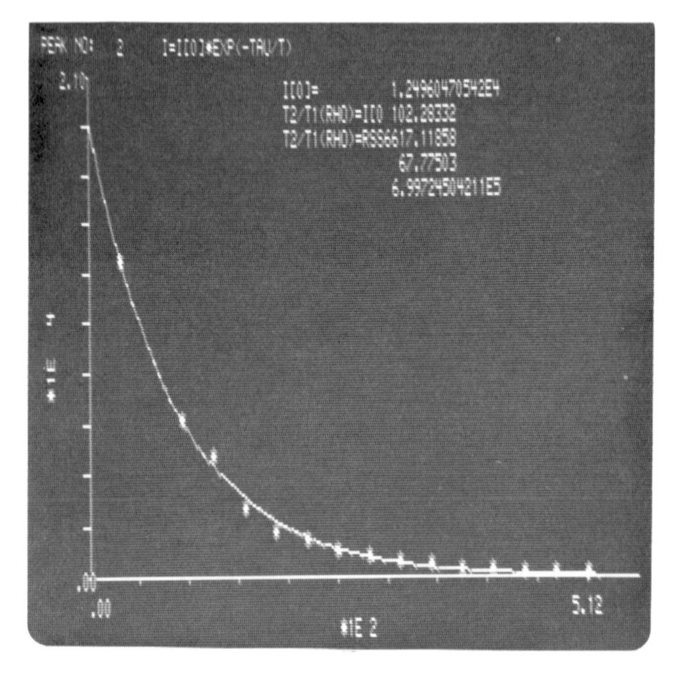

Graph 3

Separation of fats

A close look at **graph 2** reveals a less than perfect fit of a mono exponential curve. Indeed, there seems to be a marked bent around time 140 msec. This may be the effect of a diphasic sample. Fitting a diphasic model does indeed reveal two distinct curves, one with a time constant of 102 msec and another with a much shorter of 68 msec (**graph 3**). It must be noted that this graph is based on fat peaks only. Therefore, we can distinguish between two types of fat, the relative proportion being 12496/6617. Therefore, we could conclude that the two types of fat are present in the slice at a ratio 2 to 1. It has, however, to be recognized that even "pure" samples may produce diphasic relaxation curves. Therefore, the matter is open to investigation.

Generation of data

Both imaging and non imaging techniques will be investigated on the same animals and indeed on the same slices. To reduce the problems of data handling the standard procedures of imaging are being modified as such, that after the imaging sequence integral T_2 values will be obtained. The vector of 64 data points generated from the integral T_2 measurement will be appended to the image data matrices, thus allowing joint evaluation.

Perspectives

It seems that magnetic resonance imaging is a powerful research tool but because of its costs restricted to the research area. It will help to improve auxiliary traits and to gain insight into growth processes. Integral T_2 measurements hold promise of a fast and yet precise prediction of body composition. Furthermore, investigations into different classes of fat seem possible. Since the measurement time is very short no problems of artefacts due to breathing or more general movement are expected.

Looking further into the future, it is conceivable, that less costly non imaging machines can be developed, which could be used in the field for performance test or in the abattoirs for carcass evaluation. Since measurement time is below one second for each "shot", no anaesthesia might be necessary one day in the evaluation of live animals. Accuracy of prediction can be increased by taking many measurements along the body.

APPLICATIONS

Assessment of body composition by MR measurements (Stage I)

Growth curves, changes of proportion and the development of tissues are of vital interest for animal science and production of meat. The possible influence on the incidents of growth by changes of the genotype or environmental conditions, like feeding or sex differences, are the most important questions in production technique. Within the body and between tissues different growth rates can be found, these developments depend on the function of tissues or parts of tissues, respectively, like e.g. individual muscles (DAVIES 1984).

Results from growth studies had to be obtained by slaughter and dissection of the animals. So far no within animal estimates of growth curves could be derived. Applying a technique of non-invasive MR measurements can give detailed information about body composition. Repeated scannings on the growing animal provide a prediction of tissue volumes on a within animal basis.

Stage I of an experimental approach has to be the investigation of the utility of MR data for quantitative information.

The assessment of body composition will be achieved by successive images and integral T_2 curves for an optimum number of slices (scans) on living pigs.

For the initial experiment 50 pigs and their carcass sides will be used. Three liveweight classes have been chosen (20, 50, 100 kg). Up to 100 slices per animal will be recorded. The slice thickness (5 to 10 mm) will depend on the strength of the signal received to get an image of the transversal section with best contrasts to differentiate the tissues. Because of the large number of slices only FAST sequences seem to be appropriate (11 sec./scan, 2 averages) to reduce anaesthesia stress of the pigs.

The body will be separately looked at in four main components:

MUSCLE: volumetric measurements of selected muscles in the shoulder (M. supraspinatus), loin (M. long. dorsi, M. psoas major), belly and ham (M. semimembranosus, M. biceps femoris)

FAT: volumetric measurements for subcutaneous, intermuscular and kidney fat
fat water separation technique for information on the amount of intramuscular fat

BONE: selective bones of the limbs (humerus, radius and ulna, femur, tibia and fibula) and rump skeleton to be analysed for longitudinal and possibly three dimensinal growth

VISCERA: volumetric measurements for liver, kidneys and intestines, if possible heart (with ECG triggering)

Pigs will be slaughtered after scanning, and one half (carcass side) scanned again with the same number of slices as on the live animal. Anatomical dissection into muscle, fat and bone will follow to scrutinize the results of MR measurements. In STAGE I of the experiment dissection values will have to serve as reference. The 100 kg group will be scanned three times (20, 50 and 100 kg). MR data can be compared at diffenrent stages of growth.

The signals will vary (T_2 and proton density), because of the change in chemical composition. So analysis of lean and fat for water, protein and fat content of the tissues will also be carried out for each animal or its carcass in this trial.

Prediction of lean to fat ratio

Presently, lean to fat ratio provides the most important informations for assessing the value of a meat producing animal or its carcass, respectively. To reduce costs the most informative slices should be selected by MR images, either on live animals, e.g. potential breeders, or for carcass grading.

A minimum number of slices or the "optimal slice" for best prediction of lean to fat ratio should be found by calculating regressions.

Investigate improved auxiliary traits

The estimation of carcass composition by linear measurements of backfat thickness has been developed to the use of intrascopes and grading probes, progressing to ultrasonic scanning devices. The prediction accuracy of lean meat percentage in instrumental grading has r values of nearly 0.8, and a residual standard deviation of about 2.5%. The linear measures can be scrutinized and improved by MR mesurements, because MR data matrices allow reproduction of linear and two dimensional measures in any plane. This may help to correct present auxiliary traits like determination of backfat thickness by means mentioned above. Approval of newly developed grading devices could be another application of MR measurements. Prediction accuracy could possibly be improved.

MR signals without images

Methods using MR signals without imaging have been described already. These techniques could be developed for the assessment of body composition (lean to fat ratio), e.g. in the slaughter line where results have to be obtained within 10 to 20 seconds. Multi-exponential decay curves could be an approach when investigated for their predictive value by dissection of slices and chemical analysis.

Growth studies (STAGE II)

Once MR data have proven to give accurate in vivo information of tissue alterations, specific questions for growth studies will be worked on.

Individual growth curves will be derived. Effects of energy density in diets on the deposition of protein and fat at different stages of growth in any parts of the body, effects of feed additives and all kinds of growth promoting agents, like anabolic steroids or recombinant growth hormones, will be interesting subjects.

The potential of MR technique (spectroscopy) for investigating biochemical processes provides informations on the development of meat quality in vivo.

REFERENCE

DAVIES, A.S. (1984) Wachstumsverlauf von Muskeln und Knochen bei Schweinen unterschiedlicher Endgröße Vet. Diss. Hannover

CURRENT AND POSSIBLE FUTURE APPLICATION OF IN VIVO ASSESSMENT IN SHEEP BREEDING PROGRAMMES

G Simm

Edinburgh School of Agriculture
West Mains Road, Edinburgh, United Kingdom

ABSTRACT

Overfatness of lamb carcasses is a problem in many countries. Genetic improvement of carcass composition, particularly in specialised terminal sire (meat) breeds, should lead to permanent, cumulative, and cost-effective improvements. In the past, selection in these breeds has been mainly on liveweight. However, the use of *in vivo* estimates of carcass composition is expected to lead to higher rates of progress. Index selection on liveweight and ultrasonic measurements, with relative economic values of +3 for lean and -1 for fat, is expected to give annual responses of +194 g lean and +67 g fat - about 94 per cent of the response in lean, but only 26 per cent of the response in fat expected from selection on liveweight alone. Index selection on live weight and *in vivo* measurements with perfect precision is expected to give annual responses of +262 g lean and -16 g fat (a 60 per cent improvement in genetic gain in overall economic merit, compared to selection on ultrasonic measurements). Research to improve *in vivo* estimation of carcass composition for sheep breeding programmes is therefore clearly justified.

INTRODUCTION

In many countries sheep fat is produced in excess of consumer requirements. For example, Kempster, Cook and Grantley-Smith (1986) estimated that in Britain in 1984 the average lamb carcass had 20 per cent excess fat. Fairly rapid improvements in leanness could be achieved by slaughtering animals at lighter weights, or by ceasing castration of male animals. Pharmacological agents might also be used to manipulate carcass composition, though such techniques might be unacceptable to consumers in future. Also, the protein:energy ratio of the feed might be increased to improve carcass composition, though in extensive production systems this might be impractical. Although somewhat slower than these other techniques for improving carcass composition, genetic improvement by breed substitution, or within-breed selection, is potentially very useful. Genetic improvement is particularly attractive because it is permanent, cumulative and usually highly cost-beneficial. The purpose of this paper is to examine the opportunities for genetic improvement in carcass composition of sheep, and to examine the role of *in vivo* assessment of carcass composition in such programmes, and particularly how improvements in the precision of *in vivo* estimates might help in breeding programmes.

OPPORTUNITIES FOR GENETIC IMPROVEMENT IN CARCASS COMPOSITION

Breeding goals in farm animals have turned full circle over the last few centuries. Old drawings and more recent objective analyses show that "primitive" types of livestock were relatively lean. For example, Table 1 shows that the unimproved Soay breed of sheep is smaller, and has a higher proportion of muscle and bone and a lower proportion of fat than more recently developed breeds. However, from the 18th century until the earlier part of this century animals which we would now regard as excessively fat were prized and actively selected for. In the last few decades the emphasis has switched back towards leaner animals.

TABLE 1. Carcass composition of three breeds of sheep at approximately 56% of their mature live weight (McClelland et al, 1976).

	SOAY	SOUTHDOWN	OXFORD
Carcass wt. (kg)	5.7	14.4	26.5
Muscle %	59.7	54.7	50.1
Fat %	16.5	29.5	33.1
Bone %	21.9	14.8	16.4

Generally, as animals grow towards their mature liveweight, the weight and proportion of fat in the carcass increases (see Figure 1). Whilst there is still much variation between "modern" breeds in fatness at a given weight or age, this variation is reduced when breeds are compared at a common degree of maturity in liveweight. Other things being equal, leaner lambs could be produced, at a given carcass weight, by selecting breeds or sires with higher mature size, since their progeny would tend to be less mature, and hence leaner, at this given weight than progeny of smaller breeds or sires.

Breed comparisons are usually made on a representative sample of animals to enable future selection amongst breeds. That is, their purpose is to provide information for use in future selection, and not usually immediate selection amongst the animals involved in the trial. Thus, it is perfectly feasible to measure mature size and dissected carcass composition at given degrees of maturity, or specific weights or ages, in a breed comparison. However, mature size and dissected carcass composition

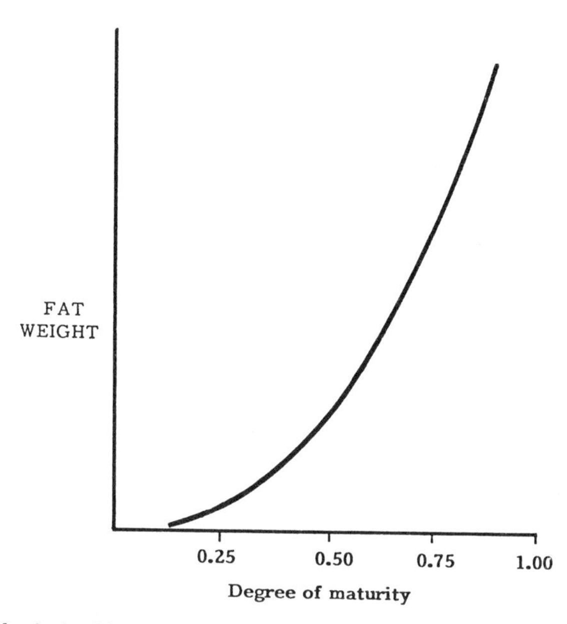

Fig. 1 **General relationship between fatness and degree of maturity in live weight**

are less practicable as selection criteria within breeds, since the measurements need to be made on candidates for selection, or their relatives. Obviously mature size takes a long time to measure, and carcass dissection is destructive (though it could be done after semen collection, or possibly in future, on identical animals derived by embryo splitting or cloning). The heritabilities of growth and carcass composition in sheep are moderately high (about 0.3, Simm *et al*, 1987b) and so selection based on an individual animal's performance is expected to give the most rapid rates of genetic improvement, if sufficiently precise *in vivo* techniques are available. Selection on immature liveweights, together with *in vivo* measurements of carcass composition is likely to be useful, both as an indirect means of increasing mature size, and for exploiting any additional genetic variation in carcass composition at a given degree of maturity (though the degree of maturity is not known at the time of selection).

IN VIVO ASSESSMENT OF CARCASS COMPOSITION IN SHEEP

The techniques used, and the problems of *in vivo* assessment of carcass composition in sheep have been reviewed by Bech Andersen (1982), Alliston (1983), Lister (1984) and Simm (1987). Probably the most widely tested method for *in vivo* assessment in sheep is pulse-echo ultrasonic measurements of fat and muscle. In general, ultrasonic measurements in sheep have given less precise predictions of carcass composition than that achieved in pigs and, to a lesser extent, in cattle. This is probably partly due to the dependence of pulse-echo ultrasonic techniques on the amount of subcutaneous fat in the animal's body. A lower proportion of the total carcass fat of sheep and cattle is in the subcutaneous depot, compared to that in pigs (see Table 2), and obviously sheep are smaller in absolute terms, so the total depth of subcutaneous fat is lower.

TABLE 2. Typical proportions of lean and proportions of fat from different depots in the carcasses of sheep, cattle and pigs (from Kempster et al, 1986).

	Sheep	Cattle	Pigs
	Proportion of tissue in carcass*		
Lean	56.1	60.3	52.1
Subcutaneous fat	11.5	7.9	16.1
Intermuscular fat	11.0	12.9	5.0
Internal fat+	3.1	1.4	1.6
Total separable fat in carcass	25.6	22.2	22.8

*NB Definition of carcass varies from species to species
+ Kidney, knob and channel fat (KKCF) for sheep; cod/udder fat for cattle, flare fat for pigs

Typically, residual coefficients of variation in lean and fat percentage, after fitting liveweight and ultrasonic measurements, range from about 0.5 to 0.7 of the original coefficients of variation (see Simm, 1987). Early results on the use of X-ray Computed Tomography (CT) measurements, together with liveweight, show improved precision of predicting carcass composition (residual coefficients of variation about 0.3 of original coefficients; Sehested, 1984). Techniques such as Nuclear Magnetic Resonance (NMR) are expected to give similar or improved precision of predicting carcass composition compared to that achieved in early trials on CT.

Selection for reduced subcutaneous fatness is expected to lead to a reduction in total fatness because of the positive and moderately high genetic correlations between fat in the different depots (see Table 3). However, selection on measurements from *in vivo* techniques which measure fat in all depots is expected to lead to greater progress. This is particularly relevant in ruminants because of the higher proportion of intermuscular fat, which is more difficult and expensive to remove from the carcass than fat from the subcutaneous or internal depots.

TABLE 3. Genetic correlations between fat proportions in different depots (from Wolf et al 1981).

	Genetic correlation (±s.e.)	
Subcutaneous, intermuscular fat (%)	+0.57	±0.19
Subcutaneous, internal (%)	+0.74	±0.19
Intermuscular, internal (%)	+0.35	±0.18

Obviously the cost and practicability of different techniques varies widely, and this needs to be considered along with precision before choosing techniques for use in a breeding programme (or for other uses). Perhaps the best way to evaluate the benefits of different techniques for use in sheep breeding is to examine the potential extra genetic gain from including different measurements in the selection criterion, rather than comparing techniques on precision alone. Before examining this in more details it will be useful to look at some of the options for selection criteria to improve carcass composition.

SELECTION CRITERIA TO IMPROVE CARCASS COMPOSITION

Possible selection criteria to improve carcass composition by genetic means include:

1. Growth rate
2. Estimated carcass lean percentage
3. Estimated carcass lean weight
4. Estimated lean tissue growth rate, or
5. Economic selection indexes.

In sheep, where production is usually linked to seasonal grass growth, the main objective is to maximise production in a given time interval. For this reason and for operational reasons, selection decisions will often be made at a fixed age, rather than a fixed liveweight. At a given age there may be a positive genetic correlation between growth rate and carcass fat weight or proportion, and a negative correlation between growth rate or liveweight and carcass lean proportion. Hence, selection on growth rate may increase fatness, and selection on estimated lean proportion may reduce carcass weight, at a given age. Neither outcome is likely to be desirable in terminal sire breeds. Even with relatively precise *in vivo* estimates of carcass composition, prediction equations for lean weight tend to be dominated by liveweight (Sehested, 1984). Similarly, lean tissue growth rate, estimated as the product of growth rate, killing out and lean proportions, tends to be highly correlated with growth rate in ruminants, especially when precision of *in vivo* estimation is low (Simm *et al*, 1987a). This is a consequence of the part/whole relationship between lean weight and live weight, but also the relatively high variation in growth rate or live weight, compared to that in killing out and lean proportions. Economic selection indexes for lean meat production theoretically give optimal weightings to live weight and *in vivo* measurements, to maximise the rate of genetic change in profitability.

Simm and Dingwall (1988) have examined the expected responses from index selection on different *in vivo* measurements, with a range of relative economic values for carcass lean and fat weights. The main comparison which they made was between selection on liveweight, and selection on indexes combining measurements of live weight, ultrasonic fat depth and ultrasonic muscle depth. Although early work shows that techniques such as CT and NMR are more precise than ultrasonic measurements, there is a lack of genetic information to allow these measurements to be evaluated in a selection index. To provide a framework for examining the potential benefits from the use of such techniques, the theoretical maximum responses in carcass composition were examined also. In this case, phenotypic and genetic correlations between the hypothetical *in vivo* measurements and the corresponding carcass measurements were assumed to be one. Also, the heritabilities of the hypothetical *in vivo* measurements were assumed to be equal to those of the corresponding traits in the dissected carcass.

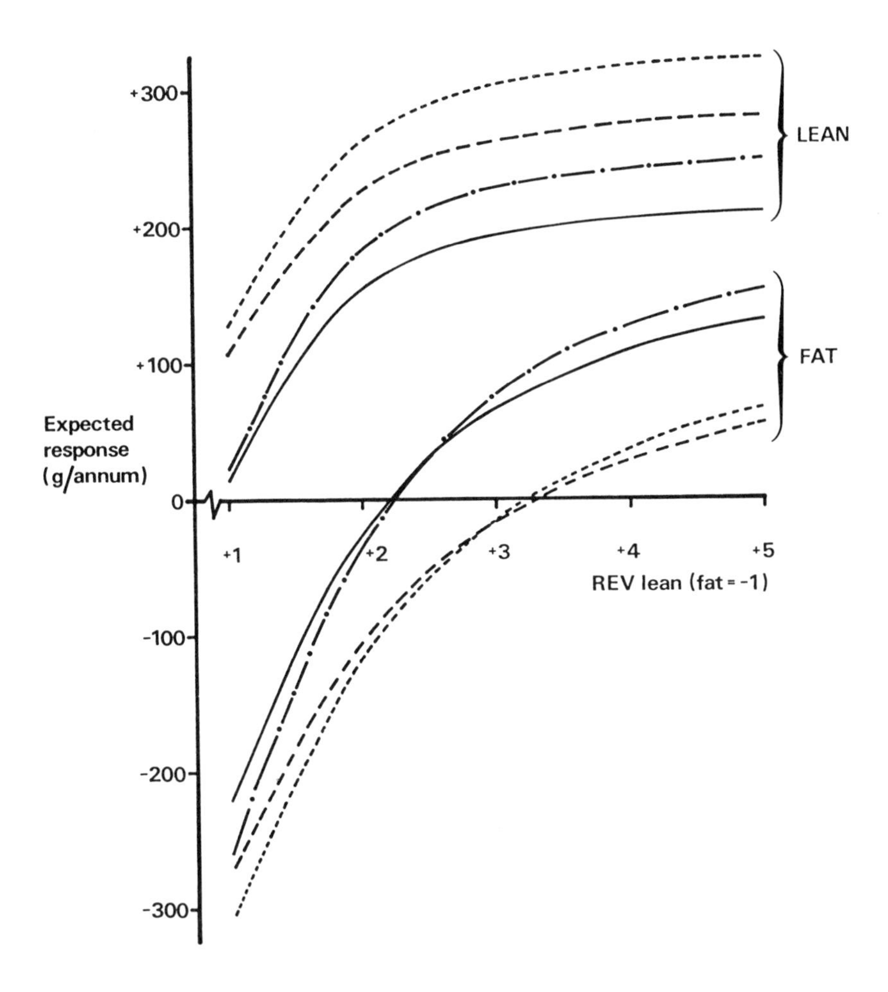

Fig. 2 Expected annual responses in carcass lean and fat weights from selection on different indices, with a range of relative economic values for lean and fat (assuming selection intensity/generation interval = 0.60).

Key:

————————	ultrasonic index, individual record
—— · —— · —— · ——	ultrasonic index, individual plus half-sib records
- - - - - - - - - - - - - - -	perfect precision, individual record
································	perfect precision, individual plus half-sib records.

TABLE 4. Expected responses in lean and fat weights from selection on different indexes (from Simm and Dingwall, 1988)

Index measurements	Expected response (g/annum)	
	Lean	**Fat**
Live weight	+206	+261
Live weight, ultrasonic fat depth	+189	+135
Live weight, ultrasonic fat and muscle depths	+194	+ 67
Live weight, perfect *in vivo* measurements	+262	- 16

+ Individual animal records only; relative economic values +3 for lean, -1 for fat.

Figure 2 and Table 4 show the most important results. The graph shows that when the relative economic value (REV) of lean is low, compared to that for fat, very low responses in lean weight are expected, whilst substantial genetic reductions in fat weight are expected. As the REV of lean increases, so the expected genetic increase in both lean and fat weights increase. Selection on an index including *in vivo* measurements with perfect precision is expected to give much higher responses in lean weight, with lower responses in fat weight than selection on an index including ultrasonic measurements.

With REVs of +3 for lean and -1 for fat, selection on an index including ultrasonic measurements is expected to give about 94 per cent of the response in lean weight, but only 26 per cent of the response in fat weight expected from selection on liveweight alone. With the same REVs, index selection with perfect *in vivo* measurement is expected to give 35 per cent higher responses in lean weight, with lower responses in fat. If these REVs are appropriate, then this extra response will lead to a 60 per cent improvement in the annual genetic change in overall economic merit.

Perhaps the benefits of increased precision of *in vivo* estimation are easier to see when responses in lean weight are compared at the same rate of change in fat weight.

For example, at zero change in fat weight, the expected annual responses in lean weight will be about 170g from individual selection on an index with ultrasonic measurements included, and about 270g from selection on an index including individual records of *in vivo* measurements with perfect precision (Simm and Dingwall, 1988).

Clearly these potential gains in response justify further research on improving precision of *in vivo* techniques in sheep. Even when new techniques appear expensive, they may be highly cost-beneficial in two stage selection programmes, after initial screening of animals with a less precise but cheaper *in vivo* technique. Also, improvements in reproductive technology such as AI and ET allow quite large improvements in the rates of genetic change achievable (eg. Smith,1986), and in the rate of dissemination of improved stock. The benefits from accelerating genetic gains through the use of embryo transfer are greatly enhanced by increasing the precision of measuring carcass composition (Steane *et al*, 1988).

CONCLUSIONS

Potentially, genetic improvement offers a permanent, cumulative, and cost-effective method of improving the carcass composition of sheep. The leanness of lambs slaughtered at a given liveweight can be increased by using a terminal sire breed with a higher mature weight than that currently used. However, further improvements can probably be made by within-breed selection of sires on immature liveweights, together with *in vivo* measurements of carcass composition. With reasonable estimates of economic values and genetic parameters, economic selection indexes combine these measurements in an optimal way. Ultrasonic measurements of fat and muscle have been quite widely tested in sheep, but early results suggest that much higher precision will be achieved with newer techniques such as X-ray CT and NMR. Index selection on ultrasonic measurements is expected to give responses of +194g lean and +67g fat (with REVs of +3 for lean and -1 for fat), while selection on an index including *in vivo* measurements with perfect precision, is expected to give responses of 262g lean and -16g fat per annum. At these REVs this extra precision is expected to give a 60 per cent improvement in the rate of genetic change in overall economic merit. Thus research for improved techniques for *in vivo* estimation of carcass composition in sheep breeding programmes is clearly justified. Although such techniques might be more expensive than ultrasonic measurements, when coupled with advances in techniques for embryo transfer, these could lead to relatively higher rates of genetic change, and thus, with effective dissemination, to high national cost benefits.

REFERENCES

Alliston, J.C. 1983. Evaluation of carcass quality in the live animal. In "Sheep Production" (Ed. W. Haresign), (Butterworths, London) pp. 75-94.

Kempster, A.J., Cook, G.L., Grantley-Smith, M. 1986. National estimates of the body composition of British cattle, sheep and pigs with special reference to trends in fatness. A review. Meat Science 17, 107-138.

Lister, D. 1984. *In vivo* measurements of body composition in meat animals. (Elsevier, London).

McClelland, T.H., Bonati, B. and Taylor, St. C.S. 1976. Breed differences in body composition of equally mature sheep. Animal Production. 23, 281-293.

Sehested, E. 1984. Computerised tomography of sheep. In "*In vivo* measurement of body composition in meat animals" (Ed. D. Lister). (Elsevier, London). pp. 67-74.

Simm, G. 1987. Carcass evaluation in sheep breeding programmes. In "New Techniques in Sheep Production". (Ed. I.F.M. Marai and J.B. Owen) (Butterworths, London). pp. 125-144.

Simm, G. and Dingwall, W.S. 1988. Selection indices for lean meat production in sheep. Livestock Production Science. (In press).

Simm, G., Smith, C. and Thompson, R. 1987a. The use of product traits such as lean growth rate as selection criteria in animal breeding. Animal Production, 45, 307-316.

Simm, G., Young, M.J., and Beatson, P.R. 1987b. An economic selection index for lean meat production in New Zealand sheep. Animal Production, 45, 465-475.

Smith, C. 1986. Use of embryo transfer in genetic improvement of sheep. Animal Production, 42, 81-88.

Steane, D.E., Simm, G. and Guy, D.R. 1988. The use of embryo transfer in terminal sire sheep breeding schemes. Proceedings of 3rd World Congress on Sheep and Beef Cattle Breeding, Paris (In press).

Wolf, B.T., Smith, C., King, J.W.B. and Nicholson, D. 1981. Genetic parameters of growth and carcass composition in crossbred lambs. Animal Production, 32, 1-7.

DISCUSSION

Chairperson: O.K. Pedersen/Denmark

I believe that we have got a tool which can be of great importance, but we have to consider some crucial problems. First, the costs for the purchase of equipment and secondly, but not less important, the costs of running the equipment. I have been informed that the running costs of the NMR-scanner in Mariensee will amount to approx. 100,000 DM per month. If the National Institute of Animal Science in Denmark should use that amount of funds it would mean that we would have to sack about 20 persons which of course would seriously affect the various research fields. If, on the other hand, we can increase our budged by the necessary amount it is a different situation which, however, seems very unlikely. Therefore, we must find other possibilities.

For smaller institutes and/or countries, such as for us in Denmark, it could be very valuable to found a close cooperation with institutes in other countries. Perhaps the EEC could be the organizing organ for such cooperation.

It must also be decided whether imaging and/or spectroscopy is needed.

It may be possible to get some companies interested in developing equipment specifically designed for research in animal production. This way we might be able to get cheaper but perfectly useful equipment.

It is of great importance that we already have a number of institutes working in this field and I have understood that it is possible for scientists from other countries to join the group set up in Mariensee.

This group will now start a basic experiment which will form the basis for further development, to study growth and influence of different nutrition on growth and development in pigs.

The paper presented by G. Simm/U.K. deals with the possibility of utilizing NMR in the selection for leanness in sheep. In Norway a program has been set up to use Computerized Tomography (CT) in pig breeding. This is of course a possibility, but in this relation one should not forget the ongoing development of statistical methods and advanced computer programs by which it is possible to handle very big amounts of data, perhaps measured with a lower certainty, but by combination of information from relatives give the basis for fast genetic improvement.

As I see the possibilities for NMR, it should mainly be used in basic research to measure basic functions in the live animal, but of course the future development and possibilities should never be ignored.

SESSION VII

SPECTROSCOPY PERSPECTIVES

Chairperson: G. Kozak-Reiss

NMR SPECTROSCOPY PERSPECTIVES

J.P. Renou, G. Monin

Station de Recherches sur la Viande
INRA Theix 63122 Ceyrat France

ABSTRACT

The future of NMR as an *in vivo* body composition tool, for imaging techniques and for spectroscopic determinations, requires to resolve some technical problems and also to establish relationships between the relevant NMR parameters and the tissue characteristics.
The development of wide bore and superconducting magnets has made up for the inherently low sensitivity of the NMR technique and the low metabolite concentrations and facilitated the emergence of this non-invasive and non-destructive technique.^{31}P NMR spectroscopy is widely used to study tissue metabolism and muscle energetics. Some very recent studies on ^{1}**H**, ^{13}**C** and ^{23}**Na** have suggested promising new applications of *in vivo* NMR.
The accurate assignment of a spectrum to a specific region within the sample is always a major problem. Surface coils provide one method of limiting the field of view of NMR spectrometers. Methods which use pulsed field gradients can now achieve a discrimination between chemical shift and spatial information.
The variation of water proton relaxation times during *post mortem* tissue ageing reflects the changes of the physical states of bound water. Localized relaxation time measurements may be a convenient technique to investigate the mechanisms of water binding in muscles.

INTRODUCTION

The combination of proton (^{1}**H**) NMR imaging and Phosporus (^{31}**P**), Carbon (^{13}**C**), and ^{1}**H** NMR spectroscopy techniques could provide a new step in the study of body composition and in the understanding of *post mortem* changes that occur in meat. Meat resarch has benefited from the development of high field NMR imaging and spectroscopy in medicine. The assesment of animal body composition likewise requires large bore magnets and chemical shift spectroscopy needs highly homogeneous fields. These parameters are technological and can be improved. On the other hand, the intrinsic characters of nuclei limit NMR studies. In the present paper there is a brief review of these experimental conditions followed by an examination of different spectroscopic methods. The major problem is now to define as well as possible the observation volume. Some methods which originate directly from imaging techniques, are very attractive. The strengths and the stabilisation of field gradient limit the applications to proton spectroscopy which benefits of small chemical shift sweep and of great sensitivity. In addition to chemical shift spectroscopy, the studies of relaxation times may determine a localization of different states of water and thus allow a better understanding of water holding capacity in meat.

EXPERIMENTAL CONSIDERATIONS

Magnets

Classical NMR spectroscopy requires a small homogeneous sample placed in a very uniform magnetic field often uniform to a part in 10^9. NMR imaging (MRI) used for large heterogeneous specimens and usually assumes that spins of all tissues precess at the same frequency. MRI however produces images of transverse sections through the body of outstanding quality. The sample size is always the most limiting factor. Magnets generate very stable fields in a diameter bore up to 1m. Permanent magnets compared with supraconducting magnets have the advantage of being less expensive, of having no running costs and a very small stray field. They are however very heavy and weights increase roughly with the magnetic field. This factor limits their use and runs counter to the current trend to work at higher field strengths. An increase in field strength yields improvements in resolution and also in the detection of nuclei with poor sensitivity, though the proton is by far the most commonly used nucleus to study living systems. The instrumental problem could be resolved by new materials such as supraconductors at a higher temperature (liquid nitrogen).

Nuclei

NMR spectroscopy is widely used to follow muscle energetic metabolism by ^{31}P NMR, or glycogen metabolism by ^{13}C NMR. Many **^{31}P** studies are made with this technique (see review by Monin and Renou, 1988),whereas metabolism has been much less studied in living systems by ^{13}C NMR.

The **^{13}C** natural abundance level restricts the number of compounds which can be detected without long acquisition times. Spectra only from lipids and glycogen at natural abundance have been obtained (Shulman, 1986). Many studies were restricted to following ^{13}C fully labelled metabolites through the various metabolic pathways of glycolysis, gluconeogenesis and the Krebs cycle or to studying the control of fluxes through these major pathways (Barany and Arus, 1986). Gochin and Pines (1985) reported an experiment which allows heteronuclear spectra to be recorded from a heterogeneous sample with the use of a surface coil. ^{13}C and ^{1}H spins evolve together to yield a ^{13}C spectrum with the satellite proton sensitivity and independent of static field inhomogeneity. The background water does not interfere, since it has no heteronuclear coupling.

Indeed, the large signal arising from tissue water impairs the study of tissue metabolism by high-resolution **^{1}H** NMR. It is desirable to use ^{1}H spectroscopy because the sensitivity of proton detection is much higher than that of ^{31}P or ^{13}C. The ^{1}H signal for an equal number of spins is 15 and 6 000 times stronger respectively than the ^{31}P and ^{13}C signal from a non-labelled metabolite. The ^{1}H method allows the detection of lower metabolite concentrations. The decrease in acquisition time makes it possible to observe fast events while saving the high signal to noise ratio. Pulse sequences (Plateau and Guéron, 1982; Hore, 1983) now achieve excellent suppression of water signal. More recently their combination with a Hahn spin-echo suppressed also the broad lipid components which hide the CH_3 lactate resonance (Heterington *et al.*, 1985; Blondet *et al.*,1986; Blondet *et al.*, 1987; Von Kienlin *et al.*, 1988). ^1H NMR spectroscopy is therefore a suitable method for studying the changes in lactate level and in intracellular pH (Seo *et al.*, 1983).

^{23}Na NMR is a convenient and rapid method of discrimination between intracellular and extracellular sodium ions by the use of extracellulary localized paramagnetic shift reagent (Gupta and Gupta, 1982). A sizeable pool of intracellular sodium ions may be NMR-invisible due to the broadening of its resonance via nuclear quadrupolar interactions (Springer C.S. Jr., 1987). NMR may be a unique noninvasive technique in studies of the role of Na^+ ions in cell regulation. Pekar *et al.* (1987) used the coherence transfer NMR to improve the sensitivity and thus detected selectively the intracellular sodium.

HIGH RESOLUTION SPECTROSCOPY

High resolution NMR spectroscopy uses different techniques for *in vivo* and *ex vivo* measurements.

The *ex vivo* measurements have many advantages: they require small samples and resolve three problems which are spatial localization, field homogeneity and magnet size. The studies are similar to classical NMR spectroscopy. The drawback lies in the preparation of samples which must be kept alive in the magnet probe. However all regulation systems present *in vivo* are not maintained.

The *in vivo* measurements use surface coils extensively to achieve spatial localization of NMR signals in animals and humans (Ackerman *et al.*, 1980). The first problem lies in the field of view. Radio frequency (RF) field intensity that varies with spatial coordinate produces spatial localization. The sensitive region has the same diameter as the coil and a thickness equal to the coil radius. The spatial localization is further improved by using pulse sequences as "depth pulse"" (Bendall, 1983). Unlike the metabolism of depth organs, the field of view problem is less important in the studies of muscular metabolism. The uniform distribution of metabolites in the tissue facilitates measurements.

LOCALIZED HIGH RESOLUTION SPECTROSCOPY

Two main methods can achieve localized high resolution spectroscopy. They differ by the use or not of pulsed gradients. The rotating frame experiment with surface coils does not need pulsed gradients but its use is limited since the penetration depth is defined by the coil radius.

The rotating frame experiments

The spatial dependence of the surface coil RF field is essential for the execution of a class metabolite mapping experiments based on the rotating frame experiment. The resonance intensity varies in an approximate sinusoidal way with the RF pulse duration (t_1). The frequency of the amplitude modulation depends on the spatial location of the nucleus giving rise to the resonance signal. Each FID retains the chemical shift information in detection period-t_2-. By Fourier transform with respect to t_1 and t_2 a two-dimensional spectrum comprising the metabolite map is obtained. Despite these attractive features, there are several potential problems with the rotating frame metabolite mapping experiment. The mapping is non-linear with distance from the coil. Intensity corrections have to be applied because of the decrease in RF strength, the spatially dependent change in slice volume, and T_1 discrimination effect. A short repetition time relative to the metabolite T_1 values leads to a degradation in resolution by smearing in the mapping dimension (Garwood *et al.*, 1984).

Gradient pulsed methods

Selected volume spectroscopy and chemical-shift imaging are two methods which use pulsed gradient sequences to define the selected volume of interest. Methods using pulsed gradients should differentiate chemical shift and spatial information. The high resolution NMR spectroscopy depends on three factors:

- the magnitude of the chemical shift itself measured in Hz
- the strength of the gradient
- the sensitivity of the nucleus observed.

The gradient strength has to be large to obtain a sufficient spatial resolution. It is worth noting that frequency shift between pixel is expressed in Hz. Typically field gradients in use are in the order of 0.5 gauss /cm which over a field of view of 50cm corresponds to ± 12.5 gauss i.e 50 000 Hz for ^{1}H. If the image has 256 points, the frequency between each pixel is about 200 Hz for ^{1}H. The ^{1}H spectrum is roughly composed of the signals of the water protons and of triglycerides CH_2 protons. The chemical shift difference between water and fat signals is 3.5 ppm which at 2T corresponds to 250 Hz. This value is the same frequency between each pixel. Lipid components could then end up overlapping water tissue at interfaces causing a ghost in the image. Increasing the field gradient strength reduces the pixel shift. A gradient much greater than the chemical shift sweep has to be used. For carbon spectroscopy the larger involved chemical shifts require the use of much too large field gradients. It corresponds respectively at 6 gauss per cm at 2T since the chemical shift sweep for ^{13}C is 200 ppm. It is useful to combine the chemical shift imaging with indirect detection sequences (Freeman *et al.*,1980). By use of this technique only protons coupled with ^{13}C will be detected and then allow a high spatial resolution.

The voxel size tends to be small. The result is that the signal to noise ratio decreases. This fact limits the application of the chemical-shift imaging to proton studies. The very low concentrations of metabolites with respect to water is hindrance even with ^{1}H sensitivity. For example the lactate signal is about 3 000 fold smaller than the water signal. As a result the voxel has to be 3 000 times larger in order to obtain an equivalent signal to noise ratio. The spatial resolution which is 1mm for water becomes 17mm ($\approx[3\,000]^{1/3}$) for lactate.

In the chemical shift imaging each pixel contains chemical shift information. The data can be displayed as a specific chemical shift mapping or a spectrum of each pixel. This technique has also been used with a surface coil placed on a human leg (Ordidge R.J. *et al.*, 1986).

LOCALIZED RELAXATION TIME MEASUREMENTS

NMR parameters are the chemical shift and the relaxation times. The latter occur with observable proton density to visualize the tissues in proton NMR imaging. The relaxation times T_1 and T_2 of tissues have a great importance by affecting the image contrast. Most NMR studies of water in biological systems use NMR relaxation measurements. The variation of water proton relaxation times during *post mortem* tissue ageing reflects the changes of the physical states of bound water. They may also be the first parameters to discriminate the meat traits. Indeed highly significant relationships are found between NMR relaxation times and some meat characteristics (Renou *et al.*, 1985; Borowiak *et al.*, 1986; Fjelkner-Modig and Tornberg, 1986.). The variation in water proton relaxation times during *post mortem* tissue ageing reflects the changes in the physical states of bound water but also gives a false image of water dynamics because of the NMR time scale (Kröker and Henkelman, 1986). Different water proton lattices are monitored by the NMR relaxation measurements and several models were proposed to explain this compartmentation. Nothing so far has proved that compartmentation corresponds to intra- and extracellular domains. Structural micro-heterogeneities were shown to be sufficient to explain the non-exponential relaxation (Wynne-Jones *et al.*, 1981). Further investigations should now focus on identifying these water lattices. Jensen *et al.* demonstrated in 1986 a technique using volume selective excitation for localized T_1 measurements. Since the chemical shift information is preserved, T_1 of different chemically shifted spectral lines can be determined separately from the localized volume. NMR micro-imaging allows a spatial resolution of up to 10µm with 2.5mm samples and thus is able to differentiate the two water lattices whenever they exist. T_2-weighted NMR images indicate the presence of two or more distinct, spatially separated classes of protons (Rothwell and Gentempo, 1985). In the *in vivo* experiment, separate images of diffusion and perfusion have been obtained (Le Bihan *et al.*, 1986). The water motions in muscular structure will be thus discriminated according to the isotrope or anisotrope nature. The few minutes of experimental time will perhaps limit the study to slow dynamic processes though new methods like the flash technique (Feinberg *et al.*, 1985) can achieve flow measurements.

CONCLUSION

NMR is not a sensitive method and moreover the metabolite concentrations are usually low. The high resolution spectroscopy of animal tissues has thus its own limitation . Therefore the ultimate goal, which is to obtain images with high resolution NMR spectra for each pixel, is far from being reached. It seems that imaging and localized spectroscpy become very complementary methods. The NMR techniques of high resolution spectroscopy which are used in chemistry should be applied to biological studies. The combination of gradient field sequences with specific pulse sequences is the most promising future for NMR investigation in the biological systems. There are large application fields in studies both of metabolism in live animals and of meat technological traits.

REFERENCES

Ackerman J. J. H., Grove T. H., Wong G. G., Gadian D. G. and Radda G.K. 1980. Mapping of metabolites in whole animals by 31P NMR using surface coils. Nature, **238**, 167-170.

Barany M. and Arus C. 1986. Lactic acid production in intact muscle, as followed by C-13 and H-1 NMR.In "Human Muscle Power" (Ed. N.L. Jones, N. McCartney, A.J. McComas). (Human Kinetics Publishers, Inc. Champaign).

Bendall M. R. 1983. Portable NMR sample localization method using inhomogeneous RF irradiation coil. Chem. Phys. Lett., **99**, 310-315.

Blondet P., Decorps M., Albrand J.P., Benabib A.L.and Remy C. 1986. Water-supressing pulse sequence for *in vivo* H-1 NMR spectroscopy with surface coils. J. Magn. Reson. **69**, 403-409.

Blondet P., Albrand J.P.,Von Kienlin M., Decorps M.and Lavanchy N. 1987. Use of rotating phase Dante pulses for *in vivo* proton NMR spectral editing with a single irradiation facility. J. Magn. Reson. **71**, 342-346.

Borowiak P., Adanski J., Olszewski K. and Bueko J. 1986. The identification of normal and watery pork by pulsed Nuclear Magnetic Resonance measurements. 32nd Europ. Meeting of Meat Research Workers, Ghent, 9-13,467.

Feinberg D.A., Crooks L.E., Sheldon P., Hönninger III J., Watts J., and Arakawa M. 1985. Magnetic Resonance Imaging the velocity vector components of fluid flow. Magn. Res. Med. **2**, 555-566.

Fjelkner-Modig S. and Tornberg E. 1986. Water distribution in porcine *M. longissimus dorsi* in relation to sensory properties. Meat Sci., **17**, 213-231.

Freeman R., Mareci T.H., Morris G.A.,.1980 Weak satellite signals in High-Resolution NMR spectra : separating the wheat from the chaff. J. Magn. Reson. **42**, 341-345.

Garwood M., Schleich T., Matson G.B., and Acosta G., 1984. Spatial localization of tissue metabolites by P-31 NMR rotating-frame zeugmatography. J. Magn. Reson. **60**, 268-279.

Gochin M. and Pines A. 1985. High-resolution NMR with a surface coil. J. Am. Chem. Soc. **107**, 7193-7194.

Gupta R.J. and Gupta P. 1982. Direct observation of resolved resonances from intra- and extracellular ^{23}Na ions in NMR studies of intact cells and tissues using dysprosium (III) tripolyphosphate as paramagnetic shift reagent. J. Magn. Reson., **47**, 344-350.

Heterington H.P., Avison M.J. and Shulman R.G. 1985. H-1 homonuclear editing of rat brain using semiselective pulses. Proc. Natl. Acad. Sci. USA, **82**, 3115-3118.

Hore P.J. 1983. Solvent suppression in Fourier transform nuclear magnetic resonance. J. Magn. Reson. **55**, 283-300.

Jensen D.J., Delayre J.L. and Narayana P.A., 1986. Localized T_1 measurements using volume selective excitation. J. Magn. Reson. **69**, 552-558.-

Kröker R.M. and Henkelman R.M. 1986. Analysis of biological NMR relaxation data with continuous distributions of relaxation times. J. Magn. Reson. **69**, 218-235.

Le Bihan D., Breton E. and Guéron M. 1986. *In vivo* Magnetic Resonance Imaging of intra-voxel incoherent motions and possibilities for their discrimination. 12th International conference on Magnetic Resonance in Biological Systems Todmoos, P256

Monin G. and Renou J.P. 1988. Spectroscopy and meat quality. CEC Seminar: The application of NMR techniques to the body composition of live animals. Mariensee, june14-15.

Ordidge R.J., Connelly A. and Lohmann J.A.B. 1986. Image-selected *in vivo* spectroscopy (ISIS). A new technique for spatially selective NMR spectroscopy. J. Magn. Reson. **66**, 283-294.

Pekar J., Renshaw P.F. and Leigh J.S. Jr. 1987. Selective detection of intracellular sodium by coherence-transfer NMR. J. Magn. Reson. **72**, 159-161.

Plateau P.and Guéron M. 1982. Exchangeable proton NMR without base-line distorsion, using new strong-pulse sequences. J. Am. Chem. Soc. **104**, 7310-7311.

Renou J.P., Monin G. and Sellier P. 1985. Nuclear Magnetic Resonance measurements on pork of various qualities. Meat Sci., **15**, 225-233.

Rothwell W.P. and Gentempo P.P. 1985. Nonmedical applications of NMR Imaging. Bruker report, **1**, 46-51.

Seo Y., Yoshizaki K.and Morimoto T. 1983.A H-1 NMR study on lactate and intracellular pH in frog muscle. Japn. J. Physiol., **33**, 721-731.

Shulman R.G. 1986.High resolution H-1 and C-13 NMR studies *in vivo*. 12th International Conference on Magnetic Resonance in Biological Systems Todmoos, L36.

Springer C.S. Jr. 1987. Measurement of metal cation compartmentalization in tissue by high resolution metal cation NMR. Ann. Rev. Biophys. Biophys. chem. **16**, 375-399.

Von Kienlin M., Albrand J.P., Authier B., Blondet P., Lotito S.and Decorps M. 1988. Spectral editing *in vivo* by homonuclear polarization transfer. J. Magn. Reson., (in press).

Wynne-Jones S., Jones D. V., Derbyshire W., Lillford P. J., Rodgers G. and Miles C.A. 1981. Water proton spin relaxation in muscle tissue. Bull. Magn. Reson. **2**, 408.

GENERAL CONCLUSIONS

E. Kallweit/FRG

It was only some thirty years ago, that preparative organic chemists began to exploit the measurement of nuclear magnetic resonance. The title of this interdisciplinary meeting marks yet another application in a rapidly growing scope, with more and more sophisticated equipment.

After Dr. Ganssen's excellent introductory lecture on the principles of NMR techniques, Drs. Heintz, Frahm, Kozak-Reiss, Offermann and Norris deliniated the state of the art in medicine, ranging from local spectroscopy, analysis of body fluids and circulation control to imaging with excellent soft tissue contrast, in absence of any known biological hazard.

Before proceeding to the discussions of NMR in animal science, Drs. Vangen and Davies reported on X-ray computerized tomography and Dr. Busk weighed the potentials of ultrasound-imaging. While X-ray CT will remain in the domain of research, the availability of advanced and reasonably prized ultrasound equipment should fill many demands in breeding practice.

In lectures by Drs. Foster, Monin, Henning, Simm and the perspectives of spectroscopy outlined by Dr. Renou emphasized the advantages animal science can draw from NMR techniques. They do not rest at the whole body imaging but do extend to spectroscopic analyses of muscle physiology from which characteristic parameters of halothane genotypes could be derived, to the assessment of bound water in muscle and to the nature of fat deposits, to name only a few.

There is only one drawback to a general application - the exorbitant cost of the equipment. An ongoing experiment at Mariensee might assist in lowering this impasse. In this experiment, imaging is replaced by the numerical assessment of tissue (Dr. Groeneveld). This is essential for the use of the technique in animal science. Instruments tailored to this requirement could be considerably less expensive in future.

The lectures, the lively discussion and the many mutual approaches for a continuing cooperation have been a rewarding experience. It gives me a great pleasure to acknowledge the contributions of all participants to a very fruitful meeting.

LIST OF PARTICIPANTS

AUSTRALIA

Dr. J. Thompson
University of New England
Department of Animal Science
Armidale, NSW 2351

AUSTRIA

Prof. Dr. W. Schleger
Institut für Tierzucht und Genetik
der Veterinärmedizin. Universität
Linke Bahngasse 11
1030 Wien

BELGIUM

Ir. M. Casteels
Rijksstation voor Veevoeding
Scheldeweg 68
9231 Melle-Gontrode

DENMARK

Dr. H. Busk
National Institute of
Animal Science
Postboks 39
8833 Orum Sonderlyng

Dr. O.K. Pedersen
National Institute of
Animal Science
Postboks 39
8833 Orum Sonderlyng

FEDERAL REPUBLIC OF GERMANY

Dr. U. Baulain
Institut für Tierzucht
und Tierverhalten FAL
Mariensee
3057 Neustadt 1

F. Düring
Institut für Tierzucht
und Tierhaltung
Olshausenstr.40
2300 Kiel 1

Dr.J. Frahm
Max-Planck-Institut für
Biophysikalische Chemie
3400 Göttingen

Dr. A. Ganssen
Platenstr. 43
8520 Erlangen

W. Griep
Institut für Tierzucht
und Tierverhalten FAL
Mariensee
3057 Neustadt 1

Dr. P. Heintz
Abt. für Radiologie
Medizinische Hochschule
3000 Hannover 61

Dr. M. Henning
Institut für Tierzucht
und Tierverhalten FAL
Mariensee
3057 Neustadt 1

E. Hüster
Institut für Tierzucht
und Tierverhalten FAL
Mariensee
3057 Neustadt 1

Prof. Dr. E. Kallweit
Institut für Tierzucht
und Tierverhalten FAL
Mariensee
3057 Neustadt 1

Dr. D. Norris
Universität Bremen
Fachbereich 2
NW 2 Leobener Straße
2800 Bremen

Dr. W. Offermann
Universität Bremen
Fachbereich 2
NW 2 Leobener Straße
2800 Bremen

A. Pfau
Institut für Tierzucht
und Tierverhalten FAL
3057 Neustadt 1 - Mariensee

Dr. K. Potthast
Bundesanstalt für
Fleischforschung
E.C. Baumann-Str. 20
8650 Kulmbach/Ofr.

Prof. Dr. Dr. D. Smidt
Institut für Tierzucht
und Tierverhalten FAL
Mariensee
3057 Neustadt 1

FRANCE

Dr. Kozak-Reiss
Faculté de Médicine Paris-Sud
Dept. de Physiologie Humaine
CCML 133 Ave. de la Resistance
92350 Le Plessis Robinson

Dr.G. Monin
Station de Recherches
sur la Viande
CRZV de Theix - INRA
63122 Ceyrat

Dr.J.-P. Renou
Station de Recherches
sur la Viande
CRZV de Theix - INRA
63122 Ceyrat

Dr.A. Talmant
Station de Recherches
sur la Viande
CRZV de Theix - INRA
63122 Ceyrat

GREECE

Dr. G. Veimos
Veterinary Institute of
Infectious and Parasitic
Diseases
25 Neapoleos Street
15310 AG. Paraskevi

REPUBLIC OF IRELAND

Dr. P. Allen
AFT, National Food Centre
Dunsinea Castleknock
Dublin 15

ITALY

Dott. G. Della Casa
Istituto Sperimentale
per la Zootecnia
Via S. Gimignano, 11
Modena

Porf. Dr. V. Russo
Istituto di Allivamenti
Zootecnici
Via Fratelli Rosselli, 107
Coviolo
42100 Reggio Emilia

LUXEMBOURG

Dr. E. Wagner
Administration des Services
Techniques de l'Agriculture
16, Route d'Esch
Luxembourg

NETHERLANDS

Ir. P. Hovenier
Vakgroep Veekfokkerij
Agrar-Universität
Postbus 3338
6700 AH Wageningen

Ir. P. Sterrenburg
Instituut voor Veeteeltkundig
Onderzoek "Schoonoord"
Postbus 501
3700 AM Zeist

NEW ZEALAND

Dr. A. Davies
Massay University
Palmerston North

NORWAY

Prof. Dr. Odd Vangen
Dept. of Animal Genetics
and Breeding
Agricultural University
of Norway
1432 Aas - NLH

PORTUGAL

M.Frausto da Silva
Estacao Zootecnica Nacional
2000 Vale de Santarem

SPAIN

Dr. C. Castrillo
Dept. Produccion Animal y
Ciencia de los Alimentos
Facultad de Veterinaria
Miguel Servet, 177
50013 Zaragoza

UNITED KINGDOM

Dr. M.A. Foster
Dept. of Biomedical Physics
and Bioengineering
University of Aberdeen
Foresterhill
Aberdeen AB9 2ZD

Dr. G. Simm
The Edinburgh School
of Agriculture
Bush Estate
Penicvik
Midlothian EH26 0PY

UNITED STATES OF AMERICA

Dr. E. Groeneveld
University of Illinois
Urbana-Champaign
1207 W Gregory Drive, 126 ASL
Urbana, Il, 61801

Prof. Dr. M.D. Judge
Purdue University
Department of Animal Science
West Lafayette, Indiana 47907

Yugoslavia

G. Fazarinc
VTOZ za Veterino
Institut za Anatom.
6100 Ljubljana

C E C

Mr. J. Connell
CEC Direction General de
l'Agriculture, DG VI
200 rue de la Loi
1049 Brussels
BELGIUM